The International Library of Ethics, Law and Technology

Volume 20

More information about this series at http://www.springer.com/series/7761

Dónal P. O'Mathúna • Iñigo de Miguel Beriain
Editors

Ethics and Law for Chemical, Biological, Radiological, Nuclear & Explosive Crises

Springer

Editors
Dónal P. O'Mathúna
School of Nursing and Human Sciences
Dublin City University
Glasnevin, Dublin 9, Ireland

College of Nursing
The Ohio State University
Columbus, OH, USA

Iñigo de Miguel Beriain
University of the Basque Country, UPV/EHU
Leioa, Spain

IKERBASQUE, Basque Foundation
for Science
Bilbao, Spain

ISSN 1875-0044 ISSN 1875-0036 (electronic)
The International Library of Ethics, Law and Technology
ISBN 978-3-030-11976-8 ISBN 978-3-030-11977-5 (eBook)
https://doi.org/10.1007/978-3-030-11977-5

Library of Congress Control Number: 2019936164

This Springer imprint is published by the registered company Springer Nature Switzerland AG.
The registered company address is: Gewerbestrasse 11, 6330 Cham, Switzerland

Contents

About the Authors

Emilio José Armaza-Armaza is "Ramón y Cajal" Researcher and Associate Professor of Criminal Law, University of Deusto, Law School, Avda, Bilbao, Spain; he can be reached at emilio.armaza@deusto.es. The author's work has focused on the study of legal and ethical implications that arise from the development and application of biotechnological innovations, essentially from the perspective of criminal law. Since 2006, he is a collaborating Researcher of the Chair in Law and the Human Genome, University of the Basque Country, Spain.

Oren Asman is Executive Director, Bioethics and Law Initiative, Sackler Faculty of Medicine, Tel Aviv University, Israel, and Professor, Department of Medical Law, Sechenov First Moscow State Medical University, Russia; he can be reached asman@tauex.tau.ac.il. Oren Asman is a Practicing Lawyer, Chair of District Psychiatric Review Committees, and a Senior Academic Faculty in the Faculty of Medicine. His research focuses on ethical and legal aspects of biomedicine and led him to initiate the Bioethics and Law Initiative at Tel Aviv University.

Yechiel Michael Barilan Associate Professor of Medical Education, Sackler School of Medicine, Tel Aviv University, Tel Aviv, Israel; ymbarilan@gmail.com. Y. Michael Barilan was an academic consultant to the first Workshops in Military Medical Ethics by the International Committee of Military Medicine. He is the author of *Human Dignity, Human Rights, and Responsibility: The New Language of Global Bioethics and Biolaw* (MIT Press, 2012).

Catherine Bertrand Service d'Aide Médicale Urgente (SAMU), France, is a Civil-Military Practitioner involved at a national level in CBRNE (chemical, biological, radiological, nuclear, and explosives) countermeasures and training. She and her coauthors are members of CBRNE European projects and NATO commissions.

Iñigo de Miguel Beriain is Investigador Distinguido, University of the Basque Country, UPV/EHU, Leioa, Spain, and Ikerbasque Research Professor, Basque Foundation for Science, Bilbao, Spain; he can be reached at INIGO.DEMIGUELB@ehu.eus. Iñigo de Miguel Beriain has been working on bioethics since 1997. He participated in an EU-funded project called EDEN, more concretely in a work package which was directly focused on CBRNE incidents' ethical and legal issues. He also participates in COST Action CA15105 European Medicines Shortages Research Network which addresses supply problems to patients.

Anna Falcone is an Italian attorney, a partner at Studio Legale Falcone, Rome. She specializes in the regulation of biotechnology and the defense of human rights. She has written a large number of papers on these topics. She is also an active politician who served as the vice president of the Comitato per il No al Referendum costituzionale del 4 dicembre.

Irene Anne Jillson is Associate Professor, Georgetown University School of Medicine, Washington, DC, USA; she can be reached at iaj@georgetown.edu. Dr. Jillson's focus on bioethics includes research, teaching, mentoring of students, presenting keynote and other lectures, and assisting in the development and implementation of institutional review boards in the USA and other countries. She is currently Principal Investigator of several projects related to responsible science, technology policy, and bioethics in low-resource countries and was Co-Principal Investigator of the landmark Comprehensive Study of the Ethical, Legal and Social Issues of Biomedical and Behavioral Research and Technology.

Chamundeeswari Kuppuswamy is Senior Lecturer, Hertfordshire Law School, University of Hertfordshire, Hatfield, UK; she can be reached at c.kuppuswamy@herts.ac.uk. Dr. Chamundeeswari Kuppuswamy is a bioethicist with expertise and experience in law and ethics on new technologies; she is a former Member of the Board of the International Association of Bioethics and a Fellow of Biocentre, London. She has advised UNESCO Bioethics Committee on human reproductive cloning.

Asier Urruela Mora is Associate Professor of Criminal Law, Faculty of Law, University of Zaragoza, Zaragoza, Spain; he can be reached at asier@unizar.es. Currently, Asier Urruela Mora is Scientific Advisor at the Chair in Law and the Human Genome, University of the Basque Country, Spain. In particular, his fields of scientific interest in the area of human genome are xenotransplantations, genome and its impact on privacy including data protection, and stem cell research.

Agnieszka Nimark is Visiting Scholar, Judith Reppy Institute for Peace and Conflict Studies, Cornell University, Ithaca, NY, USA; she can be reached at an355@cornell.edu. Dr. Nimark's initial research focused on the EU civilian crisis management within the Common Foreign and Security Policy. Gradually, her interests included the EU crisis and disaster management and broader questions of global safety governance. Currently, her research focuses on risk, uncertainty, and resilience building in the context of crisis management.

Dónal P. O'Mathúna is Associate Professor, School of Nursing and Human Sciences, Dublin City University, Ireland, and College of Nursing, The Ohio State University, USA; he can be reached at donal.omathuna@dcu.ie. Dónal O'Mathúna has degrees in Pharmacy and Bioethics. His research interests are in systematic reviewing and the ethics of disasters, particularly around disaster research ethics and humanitarian healthcare ethics.

Andrea Perin is Professor of Criminal Law, Catholic University of the North, Antofagasta School of Law, Antofagasta, Chile; he can be reached at andrea.perin@unab.cl. Andrea Perin has studied Law in Italy (University of Trento) and Spain (Chair in Law and the Human Genome, University of the Basque Country). His main academic interests are criminal law, biolaw, and science and technology studies, having a particular interest in research criminal negligence, science-based construction of normative standards, and decision-making.

Miguel A. Ramiro Avilés is Associate Professor of Legal Philosophy, School of Law, University of Alcalá, Alcalá de Henares, Spain; he can be reached at miguelangel.ramiro@uah.es. Miguel A. Ramiro Avilés evaluates clinical trials since 2001 and is currently a Member of the Research Ethics Committee of the University of Alcalá. He focuses his research on the protection of the persons who participate in biomedical research and on legal literacy and access to the rights of persons with HIV.

Andrew P. Rebera is Director, AND Consulting Group, Department for Ethics, Society, and Security, Brussels, Belgium; he can be reached at rebera@and-cg.com. Dr. Rebera's main research focuses on ethical and societal challenges posed by security and technology development; his recent work has addressed ethical issues in CBRNE security and the use of emerging technologies by law enforcement agencies. Dr. Rebera has a DPhil in Philosophy from the University of Sussex.

Roger Roffey is former Research Director at the Swedish Defence Research Agency FOI in Umeå and Stockholm, Sweden; he can be reached at Roffey.roger@gmail.com. His research background includes biotechnology, BW-defense R&D, and BW/bioterrorism threat assessments. He has been Manager for FOI biodefence R&D program; Technical Expert for the Swedish MFA for the Biological and Toxin Weapons Convention (BTWC) in 1987–2001, the Australia Group; and Inspector for UNSCOM in Iraq. For 5 years, he worked on CBRN policy issues at the Swedish Ministry of Defence.

Carlos María Romeo Casabona Professor of Criminal Law, University of the Basque Country. He is a member of the European Group on Ethics and New Technologies and has participated in more than 15 EU-funded research projects, most related to medical law. He has served as a member of the Spanish National Bioethics Committee, the Spanish National Commission on Assisted Human Reproduction, and the Spanish Ethics Advisory Committee on Science and Technology. He has published more than ten books and many chapters and articles on medical law and criminal law.

Behnam Taebi is Associate Professor, Department of Values, Technology and Innovation, Faculty of Technology, Policy and Management, Delft University of Technology, The Netherlands; he can be reached at B.Taebi@tudelft.nl. Behnam Taebi's research interests are in energy ethics, nuclear ethics, responsible research and innovation (RRI), and engineering ethics. He is the Coordinating Editor of a volume on *The Ethics of Nuclear Energy* (Cambridge University Press, 2015) and Coeditor of six special issues with *Journal of Risk Research* (2015 and 2019), *Sustainability* (2019), *Science and Engineering Ethics* (2019), *International Journal of Environmental Research and Public Health* (2019), and *Risk, Hazard, & Crisis in Public Policy* (2019).

Chapter 1
Foreword: The EU Response to Major Crises: Where Do We Come from, Where Are We, Where Should We Be?

Carlos María Romeo Casabona

1.1 Introduction

Major crises often involve ethical, political and legal issues that are extremely difficult to handle. To make a good example, nuclear accidents sometimes open a catastrophic scenario which requires that someone assumes the responsibility to mend or isolate a damaged reactor. However, this is an extremely dangerous task that often causes death to some or even all those committed to address it. For example, in the case of Chernobyl, 29 firemen and employees participating in the mitigation measures died in the days and months afterward from acute radiation syndrome. Who should carry that weight? What are the boundaries of our moral duties under those circumstances? What information are we obliged to provide to people who will have to deal with such a task? These are all issues that raise fundamental moral debates. Providing a tentative response or, at least, showing the type of elements that should be considered in the discussion, is one of the main aims of this book.

Political issues are likewise not easy to address. To make another good example, let us focus on pandemic situations. As commonly known, the best strategy to face a potentially pandemic outbreak involves an intervention at its very beginning. However, this usually means that trained teams should be sent abroad. This type of decision is highly unpopular in terms of policy making, due to a simple fact: if the team is successful and the threat is effectively erased, it will be hard for the government to capitalize on it. Nevertheless, if the initiative fails, the team suffers some losses as a consequence, and the pandemic reaches the country from which the team departed, the government might be criticised for wasting valuable resources in aiding foreign countries while the national situation required all available resources.

C. M. Romeo Casabona (✉)
University of the Basque Country, UPV/EHU, Leioa, Spain
e-mail: carlosmaria.romeo@ehu.eus

D. P. O'Mathúna, I. de Miguel Beriain (eds.), *Ethics and Law for Chemical, Biological, Radiological, Nuclear & Explosive Crises*, The International Library of Ethics, Law and Technology 20, https://doi.org/10.1007/978-3-030-11977-5_1

From the legal perspective, issues are also extremely complex, due to the diverse scope of disasters and regulation. While major crises are often huge events that defy boundaries, regulation is often constricted by national sovereignty. As a consequence of this imbalance, sometimes it is impossible to adopt the type of decisions that would fit best with the situation. Of course, one might think that international conventions, treaties or, at least, ad-hoc agreements could be concluded quite fast if needed. However, experience shows that in the aftermath of a disaster, it is difficult to make rational, fair decisions. That is why an adequate legal framework able to provide answers to the most puzzling questions is a key element in major crises management, even if we sometimes forget about it. An appreciation of this undeniable truth relies on understanding the dramatic change of mentality that has taken place in the European Union (EU) context in the last 10 years. Indeed, this short contribution aims to show the U-turn embedded by the EU in the crisis management framework, both by showing the historical milestones that have brought us into it and by providing reasons that justify the change. Finally, it will focus on the issues and gaps that yet remain and we will have to be dealt with in the future.

1.2 The EU Response to Major Crises: A Historical Background

For those who are not old enough to remember, it might make sense to highlight that the EU was not originally intended to be anything other than a kind of common economic area. In fact, the foundational Treaty of Rome made no explicit mention of a common security policy or of any kind of institutional cooperation in terms of civil protection or collective crisis management. It was not, indeed, until as late as the 1970s, in the aftermath of the Seveso and the Amoco Cadiz incidents, when a sort of embryonic Civil Protection Mechanism was created in the EU. At that moment, the European Council Resolution (Council Resolution of 26 June 1978 setting up an action programme of the European Communities on the control and reduction of pollution caused by hydrocarbons discharged at sea) was passed. Its whereas included a statement that was quite revolutionary at that time: "*Whereas the authorities of the Member States the responsibility of which it is to take action in the event of pollution of the sea by hydrocarbons must have very prompt access to information on the human and material resources which can be deployed for the control of such pollution*". However, at that moment we were yet far from any kind of common policy in terms of crisis prevention and response.

The radical change continued after a ministerial meeting in Rome, in 1985. It was a milestone in the process, as EU governments formally agreed to coordinate their civil protection strategies for the first time. As a consequence, a number of preliminary initiatives that laid the foundations for today's extensive coordinated approach for dealing with and planning for major disasters were approved between 1985 and 1994, such as the Resolution of the Council and the representatives of the

Governments of the Member States, meeting within the Council of 25 June 1987 on the introduction of Community Cooperation on Civil Protection; the Resolution of the Council and the representatives of the Governments of the Member States, meeting within the Council of 13 February 1989 on the new developments in Community cooperation on civil protection; the Resolution of the Council and the representatives of the Governments of the Member States, meeting within the Council of 23 November 1990 on Community cooperation on civil protection; and the Resolution of the Council and of the representatives of the Governments of the Member States, meeting within the Council of 8 July 1991 on improving mutual aid between Member States in the event of natural or technological disaster.

The principal result of this normative development was the creation of the embryonic Civil Protection Mechanism. Soon afterwards, in 1997, the Council of the European Union took an important step forward by approving a major civil protection action programme that ran from 1 January 1998 to 31 December 1999, through the Council Decision 98/22/EC of 19 December 1997 establishing a Community action programme in the field of civil protection. This Action program *"directed the Commission to enhance its efforts aimed at the pooling of member state expertise, facilitating mutual assistance, and offering training programs. By the end of the decade, the Commission was slowly building capacity to help carry out the intentions stated in the various Council declarations- but still had very little to show in terms of actual policy successes"* (Boin et al. 2013, 638).

1.3 The EU Current Legal Framework: Towards an EU Common Legal Framework

The terrorist attacks of the early 2000s in New York, Madrid or London stressed the political need to build a common strategy to deal with natural or man-made disasters. In 2006, a study commissioned by Mr. José Manuel Barroso and Mr. Wolfgang Schüssel produced a capital Report, authorized by Michel Barnier, former French Minister for Foreign Affairs and former Member of the European Commission (Barnier 2006). It suggested 12 measures that were considered especially important in order to enforce the EU capacity to respond to a crisis situation (Ahman and Nilson 2008, 101):

- A European-wide civil protection force named Europe Aid;
- Support for the force in seven ultraperipheral regions of the European Union;
- Creation of a Civil Security Council and strengthening of the General Affairs and External Relations Council;
- A one-stop shop for humanitarian response;
- An integrated European approach to anticipate crises;
- Six EU regional delegations to specialize in crisis management;
- A clear information system for European citizens;
- Sharing of consular resources;

- Creation of flying consular teams;
- Setting up European consulates in four pilot zones;
- Drawing up a European consular code;
- Laboratories to specialize in the fight against bio-terrorism and the naming of victims.

Most of these recommendations where addressed by the legal framework designed by the Treaty of Lisbon (2007). The most remarkable innovation introduced by the Treaty in terms of major disaster situations was the so-called "Solidarity Clause", asup described by article 222 of the Consolidated version of the Treaty on the Functioning of the European Union.

Indeed, it was an extremely important tool not yet because of its practical implications (it has never been used), but because it reflected quite well the idea that defense against major disasters should be built on the basis of mutual solidarity between Member States. In the aftermath of the Treaty, the EU continued to develop a complete legal framework regarding major disasters. Two major regulations where produced in recent years: Decision No 1313/2013/EU of the European Parliament and of the Council of 17 December 2013 on a Union Civil Protection Mechanism and the EU Integrated Political Crisis Response Arrangements (IPCR). These are to be complemented by the continuous development of the EU Internal Security Strategy (ISS), as designed by the Treaty of Lisbon and the Stockholm Programme, and complementary tools such as Regulation (EU) No 514/2014 on the Asylum, Migration and Integration Fund and the instrument for financial support for police cooperation, preventing and combating crime, and crisis management.

1.4 The Issues That Yet Remain: Lessons for the Future

If we are to arrive at any conclusion, it must be this: the EU has undoubtedly advanced in the construction of a common response strategy to major crisis situations. At the present moment, the EU institutions have assumed a decisive role as coordinators of any effort made to face a situation that cannot be handled at the national level. This is an impressive improvement in the EU crisis management framework, but also significant evidence of the level of integration reached by this political body. The current situation, fortunately, is completely different to what it was merely 20 years ago and the future seems promising enough.

However, some issues yet remain as a permanent threat to this desirable scenario. First, the tension between the Member States' interest to keep their sovereignty and the need to somehow share it with the EU continues to be high. This is particularly true in cases in which national security is involved and sensitive information might be at stake. Second, we need to test and check the tools designed by the new framework, such as the IPCR. This will probably take time and will require intensive efforts. Finally, we must keep in mind that Brexit will create a situation of extreme incertitude with the United Kingdom, a country that has traditionally played a key role in terms of the EU security policies. Let us try not to compromise it, if we wish to keep on improving our collective security and our capability to react against major natural or man-made crises.

References

Ahman, Teresa, and Claess Nilson. 2008. The community mechanism for civil protection and the European Union Solidarity Fund. In *Crisis management in the European Union: Cooperation in the face of emergencies*, ed. Stephan Olsson, 83–108. Dordrecht: Springer.

Barnier, Michel. 2006. For a European civil protection force: Europe aid. http://ec.europa.eu/archives/commission_2004-2009/president/pdf/rapport_barnier_en.pdf. Accessed 27 Feb 2018.

Boin, Arjen, Magnus Ekengren, and Mark Rhinard. 2013. *The European Union as crisis manager. Patterns and prospects*. Cambridge: Cambridge University Press.

Treaty of Lisbon amending the Treaty on European Union and the Treaty establishing the European Community, signed at Lisbon, 13 December 2007. http://eur-lex.europa.eu/legal-content/EN/ALL/?uri=OJ%3AC%3A2007%3A306%3ATOC. Accessed 27 Feb 2018.

Chapter 2
Introduction to Ethical and Legal Issues in CBRNE Crises

Iñigo de Miguel Beriain and Dónal P. O'Mathúna

Abstract This chapter introduces the aims and objectives of this book. Books on chemical, biological, radiological, nuclear and explosive (CBRNE) crises are usually technical and scientific. This introduction highlights some of the reasons why a volume on the ethical and legal dimensions of CBRNE crises is needed. The chapter then provides a brief summary of each of the chapters and how these develop the topic more generally. It concludes by acknowledging the EU funding which made it possible to produce this volume.

Keywords Ethics · Law · CBRNE crises

On the night of 2–3 December 1984, an incident was reported in a chemical plant of the Union Carbide India Limited (UCIL) in Bhopal, Madhya Pradesh, India. The causes remain unclear, but its consequences are well known: thousands of people (up to 16,000 according to sources) died a terrible death, while another 38,478 suffered temporary injuries and approximately 3900 more suffered severe and permanently disabling injuries (Eckerman 2005). Many consider this the worst industrial disaster to have ever happened.

Twelve years later, on 26 April 1986, a safety test in the Chernobyl Nuclear Power Plant led instead to a massive release of radioactive material into the atmosphere. In the following weeks, 28 emergency workers ("liquidators") died from acute radiation syndrome. A report by The Chernobyl Forum (2005) calculated that

I. de Miguel Beriain
University of the Basque Country, UPV/EHU, Leioa, Spain

IKERBASQUE, Basque Foundation for Science, Bilbao, Spain
e-mail: INIGO.DEMIGUELB@ehu.eus

D. P. O'Mathúna (✉)
School of Nursing and Human Sciences, Dublin City University, Glasnevin, Dublin 9, Ireland

College of Nursing, The Ohio State University, Columbus, OH, USA
e-mail: donal.omathuna@dcu.ie

D. P. O'Mathúna, I. de Miguel Beriain (eds.), *Ethics and Law for Chemical, Biological, Radiological, Nuclear & Explosive Crises*, The International Library of Ethics, Law and Technology 20, https://doi.org/10.1007/978-3-030-11977-5_2

during the subsequent years up to 4000 people died of cancer caused by exposure to radiation. The consequences of this accident compare only to those produced at the nuclear reactor in Fukushima, Japan, after it was hit by a devastating tsunami on 11 March 2011. In the latter case, it has been estimated that this might cause an additional 130 (range, 15–1100) cancer-related mortalities and 180 (range, 24–1800) cancer-related morbidities in the years ahead (Ten Hoevea and Jacobson 2012).

As terrible as they were, all these incidents can hardly be compared with a biological crisis, such as the Black Death pandemics in the fourteenth century, which resulted in the deaths of more than 75 million people or, more recently, the 1918 Spanish flu pandemic, which caused the deaths of more than 50 million people (Anon 2013). One might, of course, consider that these types of scenarios could hardly take place at the present moment due to advances in health care and disease prevention. However, the 2009 influenza A H1N1 outbreak was declared a pandemic and the scale of its impact has recently been found to have been substantially greater than was originally presumed. In 2013, a more extensive evaluation of the pandemic found that it led to about 200,000 deaths, ten-times more than previously believed (Simonsen et al. 2013). Between 2014 and 2016, the Ebola virus claimed the lives of over 11,000 West Africans with the world's medical resources apparently unable to do anything other than watch and hope the viral outbreak would quickly run its course and fade away (World Health Organization 2018).

All the terrible accidents and natural disasters mentioned created massive response, logistical, medical and economic challenges. They also create impressive number of ethical and legal issues that require urgent attention and clear answers. In the moment of crisis, addressing such issues proved to be impressively difficult, as exemplified most clearly during the Ebola virus disease outbreak of 2014–2016. Together, such incidents are being called chemical, biological, radiological, nuclear and explosive (CBRNE) events or crises. The ethical issues in disasters more generally, and CBRNE more specifically, are only now beginning to be addressed. This can be seen in the beginnings of topics like disaster bioethics (O'Mathúna et al. 2014) and humanitarian ethics (Slim 2015). While the ethical issues themselves have been known to responders, research has shown that little careful analysis has been given to the ethical dilemmas themselves and how they can best be resolved. We are, indeed, still far from arriving at a general consensus on how to address the legal and ethical dimensions of these situations, and further analysis and reflection are urgently needed. Adding to the complexity is the need to provide high-quality evidence to guide those responding to disasters and CBRNE events. More research is being called for to provide this evidence, which will also generate research ethics issues that need to be examined and addressed (O'Mathúna 2010, 2018).

Regarding the law and CBRNE events, the situation is not any better. The European Union (EU) has only recently updated some of the tools designed to face major crisis scenarios and most of them have not been adequately evaluated yet. The Integrated Political Crisis Response (IPCR) agreements seem very promising, but their real practical utility remains unclear, while much ambiguity remains about the practical implementation of the solidarity clause by Member States. The situation regarding personal data access is of great concern. Directive 45/96/EC of the

European Parliament and the Council of 24 October 1995 on the protection of individuals with regard to the processing of personal data and on the free movement of such data, which is the EU key legal tool, has become obsolete. However, its replacement, the General Data Protection Regulation, is yet to be tested in crisis scenarios. Therefore, additional resources are needed to help identify the legal and ethical issues in CBRNE crises, develop clear ways of responding to them, and develop policies and training materials that help people and communities prepare for and respond to such events.

Last, but not least, the EU is currently suffering from an exceptional situation, the Brexit issue. As this book goes to press, it continues to remain totally unclear whether the United Kingdom (UK) and the EU will arrive at some kind of agreement on the British exit from the Union. Reaching some agreement is of the highest importance in terms of CBRNE crisis prevention and response, since the cooperation between the UK and the EU Member States on these issues has been crucial traditionally.

Similarly, tensions rising at the moment inside the Union might introduce some malfunctions in terms of cooperation and information sharing. It is hard to foresee whether the strength of the EU-sceptical political options will create a new scenario characterized by an increasing reluctance to share protected information. If this were to become the case, we would anticipate a worsening of our preparedness for major CBRNE crises.

This book begins to address the need for materials that help people examine and address ethical and legal issues on CBRNE crises, and aims to stimulate further discussion of and reflection on these crucial issues. We offer these chapters to promote discussion and deliberation, not as the final word on any of these challenging issues. The book contains chapters on both ethical and legal topics. Rather than divide the volume according to these fields, we have chosen to start with chapters that take a broad, general approach to the topic from ethical, legal and regulatory perspectives. The later chapters focus in on specific topics with particular ethical or legal concerns.

The editors are particularly grateful to Professor Carlos María Romeo Casabona for his Foreword to this volume. Relying on his extensive experience with the EU, he provides the historical background to how the EU has developed in its approach to crisis planning and management. He highlights how the political and legal approaches to major crises have developed and changed over recent decades. The introductory chapter by Dónal O'Mathúna and Iñigo de Miguel Beriain, editors of this volume, gives further background to the importance of ethics and law in CBRNE crises. We briefly review each chapter in this volume, and explain the origins of this work in two EU-funded projects.

The next chapter is written by the experienced Swedish researcher, Roger Roffey, and explores the EU as a key actor in disaster situations. To this purpose, the author overviews how the EU has developed as a crisis actor in the case of CBRNE incidents. Roffey explores the changes that have taken place within the EU in recent years. This general exploration of regulatory dimensions is followed by Andrew P. Rebera's chapter examining the ways that ethics can and should be incorporated into all aspects of CBRNE security. He argues that the best way to do this is through developing ethics guidelines, codes of conduct, and ethics training for all actors in CBRNE security. While noting some overlap with legal approaches, ethics also highlights the necessity of developing on-the-spot ethical decision-making skills. Rebera acknowledges how daunting this task may be, but challenges readers to engage in the task.

Irene Jillson introduces a broad range of approaches to ethics. Much scholarship has focused on using the four ethical principles of autonomy, beneficence, nonmaleficence and justice, but more recently a global approach to bioethics has sought to include ethical approaches from outside Western philosophical traditions. Jillson reviews some of these approaches and proposes a global health ethics framework which would be applicable to CBRNE crises. Turning to the role of governance, Behnam Taebi explores the way new technology can introduce new risks and ethical challenges. Using the example of nuclear crises, he explores how *good* risk governance attempts to identify ethical challenges and provide normative responses to them. Taebi focuses on distributive and procedural justice in nuclear disasters because of their emphasis on taking into consideration the interests of all members of society. Such an approach is highly relevant for all types of CBRNE crises.

Returning to approaches to crisis situations adopted by the EU, Agnieszka Nimark describes one of the newest political mechanisms envisaged to optimize the political response to crisis situations: the Integrated Political Crisis Response (IPCR) arrangements. These new agreements, which replace the Crisis Coordination Agreements (CCA), are meant to add more flexibility to crisis response mechanisms in the EU and strengthen cooperation between the different relevant agents in a major crisis situation. They also create useful new tools, such as the Integrated Situational Awareness and Analysis (ISAA) tool. However, their performance still needs to be fully tested, but Nimark's contribution shows that some weaknesses can already be foreseen.

The next chapter is the first to examine an important concept in this area, namely solidarity. Iñigo de Miguel Beriain provides an extensive review of the Solidarity Clause. This is an extremely important legal tool, being the only binding regulation that obliges EU Member States to support each other if circumstances overwhelm an individual state's response capacities. This is followed by a chapter written by Chamundeeswari Kuppuswamy. She begins with the observation that after disasters, many people act with solidarity, comradery and altruism. She claims that this is how it should be, and argues her case on the basis of moral duties that we all have to ourselves and others. She bases her approach on Alan Gewirth's writings, claiming that this approach to solidarity ethics should be imbedded in European policy and action as a way to fulfil human rights. Solidarity will also be applied to ethical dilemmas in clinical research, but before that Miguel Ángel Ramiro Avilés examines the possibility of performing clinical trials in CBRNE crises and the legal challenges this involves. He focuses on the importance of balancing public health and civil rights, and the difficulties involved here. He describes the need to ponder ethical principles and legal rules, when the agents causing illness or death in people, animals or plants are resistant to current medicines or have no approved treatments.

He proposes some suggestive guidelines that might serve as an effective response to such ethical and legal crossroads.

The next chapter continues the focus on clinical research ethics, and provides an example of the application of solidarity to this area. Oren Asman and Michael Barilan note that the ethical challenges of clinical research are exacerbated in CBRNE situations. Although the challenges are intensified, they argue that established bioethical values and approaches from other research settings are applicable to CBRNE research. They argue for a solidarity-informed approach to research ethics, and provide practical examples of how this could be implemented.

Research generates data, but so do many other systems, which leads to consideration of mechanisms to protect personal data in emergency situations, such as in CBRNE crises. The chapters by Asier Urruela Mora and Anna Falcone provide a paramount analysis of the legal challenges with personal data in crises. Asier Urruela Mora takes a more general approach, and then Anna Falcone narrows the discussion to Directive 95/46/EC on the protection of individuals with regard to the processing of personal data and on the free movement of such data. This regulation was replaced in 2018 by the Regulation of the European Parliament and of the Council on the protection of individuals with regard to the processing of personal data and on the free movement of such data (General Data Protection Regulation). In spite of this, Directive 95/46/EC continues to provide many of the general principles used in GDPR. Together, both chapters show how data protection regulation has evolved in the EU, identifying the weaknesses posed by the Directive and the solutions proposed by the Regulation. The authors have made a special effort to address the concrete issues that exceptional situations like CBRNE create and the possible exceptions to the general protection to individual privacy depicted by the EU regulation, a topic that continues to be relatively unexplored.

The remaining four chapters in the volume tackle specific challenges with CBRNE events. Catherine Bertrand and her colleagues examine the ethical challenges with triage. Many CBRNE crises lead to situations where insufficient resources are available to meet all the needs. The authors review the technical aspects of triage, but note that many ethical decisions are also involved. Different approaches to triage exist, but little research has been conducted to validate their effectiveness, although European projects are currently conducting such trials. Another ethical challenge for healthcare responses is the duty to care. Dónal O'Mathúna examines the origins of the ethical obligation that physicians traditionally had to response to dangerous situations, even those that put their own lives in danger. Some have questioned whether such an obligation continues to exist, although many act accordingly. Regardless, it is also recognized that employers and states have ethical responsibilities to train and support healthcare workers as they respond to CBRNE crises.

In the next chapter, Emilio José Armaza-Armaza analyses an issue that has been addressed in varying ways by different national criminal laws. Namely, Armaza addresses those situations in which a diseased person refuses to isolate him- or herself from the rest of society, thus creating risks that might extend worldwide. Such actions could cause severe harm to other people's health and contribute to the spread of pandemics to territories that had not been affected beforehand. The legal dilemmas and options are addressed in Armaza's chapter. The final chapter by Andrea Perin is devoted to identifying scientists' responsibilities in risk prevention processes. To this purpose, he uses the L'Aquila earthquake case in Italy as an illustrative example. Through this case, Perin focuses on the relationships between scientific knowledge ('regulatory science'), normative expectations, decision-making and criminal negligence for 'failed' risk assessment and management.

As noted at the beginning of this chapter, and affirmed by all the chapter authors, this book is offered as a stimulus to further discussion and debate. The challenges faced by society, political leaders, and those who respond to CBRNE crises are many and varied. Among them are many difficult ethical and legal issues that call for urgent and careful reflection. We hope that this book will go some way to stimulating such reflection, and also lead to practical guidelines and directives that will help those who will be forced to make decisions during the next crisis they face. We dedicate this book to the brave men and women who risk their lives for the good of others caught up in such terrible events.

This book would not have been possible without the time and effort of its authors. We as editors are very grateful for their contributions and scholarship. The book is the result of funding provided by the European Commission to two projects. The research project EDEN (End user driven DEmo for cbrNe; https://www.eden-security-fp7.eu/) was funded by the European Commission through its Framework VII Programme. COST Action IS1201: Disaster Bioethics was funded through Horizon 2020 and the COST Association (http://disasterbioethics.eu/). These two projects worked together to organise two workshops held at Universidad de Deusto in Bilbao, Spain. Preliminary versions of many of these chapters were presented at these workshops and benefited from feedback from the attendees. We are grateful to the Commission for the funding which allowed these workshops to take place.

References

Anon. 2013. *The five deadliest outbreaks and pandemics in history*. Robert Woods Johnson Foundation, December 13. https://www.rwjf.org/en/culture-of-health/2013/12/the_five_deadliesto.html. Accessed 27 Feb 2018.

Eckerman, Ingrid. 2005. *The Bhopal saga: Causes and consequences of the world's largest industrial disaster*. Hyderabad: Universities Press.

O'Mathúna, Dónal. 2010. Conducting research in the aftermath of disasters: Ethical considerations. *Journal of Evidence-Based Medicine* 3 (2): 65–75.

———. 2018. The dual imperative in disaster research ethics. In *SAGE handbook of qualitative research ethics*, ed. Ron Iphofen and Martin Tolich, 441–454. London: SAGE.

O'Mathúna, Dónal P., Bert Gordijn, and Mike Clarke, eds. 2014. *Disaster bioethics: Normative issues when nothing is normal*. Dordrecht: Springer.

Simonsen, Lone, Peter Spreeuwenberg, Roger Lustig, Robert J. Taylor, Douglas M. Fleming, Madelon Kroneman, Maria D. Van Kerkhove, Anthony W. Mounts, W. John Paget, and the GLaMOR Collaborating Teams. 2013. Global mortality estimates for the 2009 influenza pandemic from the GLaMOR Project: A modeling study. *PLoS Medicine* 10 (11): e1001558.

Slim, Hugo. 2015. *Humanitarian ethics: A guide to the morality of aid in war and disaster*. London: Hurst.

Ten Hoevea, John E., and Mark Z. Jacobson. 2012. Worldwide health effects of the Fukushima Daiichi nuclear accident. *Energy & Environmental Science* 5 (9): 8743–8757.

The Chernobyl Forum. 2005. *Chernobyl's legacy: Health, environmental and socio-economic impacts and recommendations to the governments of Belarus, the Russian Federation and Ukraine*, 2nd edn. https://www.iaea.org/sites/default/files/chernobyl.pdf. Accessed 27 Feb 2018.

World Health Organization. (2018). Ebola virus disease: Fact sheet. http://www.who.int/mediacentre/factsheets/fs103/en/. Accessed 27 Feb 2018.

Chapter 3
The EU as an Actor in CBRNE Crisis: A General Picture

Roger Roffey

Abstract This chapter presents an overview of how the European Union (EU) has developed as a crisis actor in the case of biological, chemical, radiological, nuclear and explosive (CBRNE) incidents. The EU's policies rely on preventing proliferation of Weapons of Mass Destruction (WMD) and CBRNE terrorism which have been developed in different policy areas. The responsibility for responding to CBRNE incidents rests with the Member States (MS). The EU has developed crisis management procedures and tools to support the MS in case of a crisis with cross-border implications. Coordination and information sharing mechanisms, capacity building, joint exercises, and sharing of best practices are examples of EU actions which support and complement the actions taken by MS (EU. EU capacities to respond to CBRN attacks and CBRN incidents. Submitted by Belgium on behalf of the European Union, 2010). In 2016, the EU presented its new Global Strategy for its foreign and security policy as a follow-on to its 2003 Security Strategy, though it has limited references to WMD (European Union. Shared vision, common action: a stronger Europe, a global strategy for the European Union's Foreign and Security Policy, Brussels, June. https://eeas.europa.eu/archives/docs/top_stories/pdf/eugs_review_web.pdf. Assessed 28 Sept 2018, 2016; Lundin L-E. The European Union and weapons of mass destruction: a follow-on to the global strategy? EU Non-Proliferation Consortium. Non-proliferation Paper. no. 58. Stockholm International Peace Research Institute. May, 2017). The EU needs to further improve preparedness, coordination and CBRNE Action Plans to better handle future CBRNE incidents.

Keywords CBRNE EU policy · CBRNE Action Plans · CBRN terrorism · CBRNE preparedness · CBRNE Centres of Excellence (CoE) · Civil protection · Health Security · WMD Non-proliferation

R. Roffey (✉)
Swedish Defence Research Agency, Stockholm, Sweden
e-mail: Roffey.Roger@gmail.com

D. P. O'Mathúna, I. de Miguel Beriain (eds.), *Ethics and Law for Chemical, Biological, Radiological, Nuclear & Explosive Crises*, The International Library of Ethics, Law and Technology 20, https://doi.org/10.1007/978-3-030-11977-5_3

Abbreviations

BTWC	Biological and Toxin Weapons Convention
CCA	EU Emergency and Crisis Coordination Arrangements
CECIS	Common Emergency Communication and Information System
CoE	Centres of Excellence
DG DEVCO	Directorate-General for International Cooperation and Development
DG ECHO	DG for Humanitarian Aid and Civil Protection
DG	Directorate-General
DRMKC	Disaster Risk Management Knowledge Centre
EC	European Commission
ECAP	European Capability Action Plan
ECTC	European Counter-Terrorism Centre
ECURIE	European Community Urgent Radiological Information Exchange
EEAS	European External Action Service
EEODN	European Explosive Ordnance Disposal Network
EERC	European Emergency Response Capacity
EMC	European Medical Corps
EOD	Explosives Ordnance Disposal
ERCC	Emergency Response Coordination Centre
ESDP	European Security and Defence Policy
ESS	European Security Strategy
EU	European Union
EUCPM	European Union Civil Protection Mechanism
EWRS	Early Warning and Response System
FBI	Federal Bureau of Investigation
FP7	Seventh Framework Programme
GHSAG	Global Health Security Action Group
GHSI	Global Health Security Initiative
GICNT	Global Initiative to Combat Nuclear Terrorism
HEOF	Health Emergency Operational Facilities
HSC	Health Security Committee
IcSP	Instrument contributing to Stability and Peace
INTCEN	EU Intelligence Analysis Centre
IPCRC	Integrated Political Crisis Response Capabilities
ISAA	Integrated Situational Awareness and Analysis
MIC	Monitoring and Information Centre
MS	Member States
PNR	Passenger Name Record
PSI	Proliferation Security Initiative
UNICRI	United Nations Interregional Crime and Justice Research Institute
RAS BICHAT	Rapid Alert System used for exchanging information due to deliberate release of chemical, biological and radio-nuclear agents

UNGA	United Nations General Assembly
UNSC	United Nations Security Council
WHO	World Health Organisation
WMD	Weapons of mass Destruction

3.1 EU Non-proliferation and WMD Strategy

The EU in 2003 adopted the European Security Strategy (ESS) which stresses that WMD proliferation, international terrorism, regional conflicts, state failure and organised crime should be addressed through ‘effective multilateralism’ (Council 2003a, 2008a). The most frightening scenario is one in which terrorist groups acquire WMD that could inflict damages on a scale previously possible only for States (Council 2005a). The 2003 EU strategy against proliferation of WMD was adopted based on a Swedish proposal (Council 2003b) and every 6 months a progress report is prepared (Council 2015a). The strategy aims to prevent third countries and terrorists from acquiring WMD-materials and their means of delivery. This will be done by making use of all available EU-instruments to prevent, deter, halt and, if possible, prevent proliferation, including implementing export control policies and international treaties (Council 2004a; Alvarez-Verdugo 2006). A WMD Monitoring Centre was created in 2006 as a coordination mechanism for non-proliferation policies in the EU (Council 2006a; Van Ham 2011). The EU also supports several UN Conventions on specific types of terrorism and promotes a Comprehensive UN Convention on International Terrorism (UNGA 2007).

With the Lisbon Treaty, the European External Action Service (EEAS) has taken over non-proliferation issues including implementation of the WMD Strategy (Council 2010a) and has strengthened the profile of the EU as an international security actor. An EEAS Principal Advisor and Special Envoy for Non-proliferation and Disarmament now exists. The EU is a key player in all the major multilateral frameworks relating to non-proliferation and WMD arms control, as well as a provider of financial and technical assistance in support of non-proliferation.

In 2005, non-proliferation clauses were inserted into agreements with third countries that request the full compliance with and national implementation of obligations under international instruments (Council 2003c). The EU’s action plan was updated in 2008 to include intensified efforts to counter proliferation flows, consider sanctions for acts of proliferation, develop measures to prevent intangible transfers of knowledge, raise awareness in scientific circles, and improve international cooperation on non-proliferation policies and export controls (Anthony and Grip 2013; Council 2008b). The EU will promote and support national implementation of related international regimes and the United Nations Security Council (UNSC) resolution 1540 (Roffey 2011; UNSC 2004, 2008).

The export control regimes cover a broad range of agents, materials and equipment that can be used to develop CBRN weapons as well as their delivery systems. Most of the covered dual-use items are fairly sophisticated and of a type required by a state clandestine program for WMD development. (Council 2009a).

There are initiatives focusing on practical international cooperation to cover new proliferation concerns like stopping shipments with WMD, such as the Proliferation Security Initiative (PSI) (Council 2004b). Another example is the Global Partnership against the Spread of Weapons and Materials of Mass Destruction that was formed to better coordinate the ongoing international non-proliferation, threat reduction programs and prevent terrorists from gaining access to WMD or materials. Another initiative is the Global Initiative to Combat Nuclear Terrorism (Anthony 2004). These initiatives do not have universal membership but engage a select like-minded number of states, mainly Western countries.

3.2 The EU Instrument Contributing to Stability and Peace, Including CBRN Centres of Excellence Initiative

The Directorate-General for International Cooperation and Development (DG DEVCO) is the main implementer of external EU non-proliferation assistance projects and acts as the single contact point for stakeholders both inside and outside the EU (Grip 2011; Council 2013). The EU Instrument contributing to Stability and Peace 2014–2020 (IcSP), formerly the Instrument for Stability, funds measures to enhance CBRNE safety/security practices at civilian facilities, strengthen effective control of illicit CBRNE trafficking, the enforcement of effective export controls on dual-use goods, and promote regional cooperation (Grip 2011; Commission 2009a).

The EU CBRN Centres of Excellence (CoE) initiative was started in 2010 and is aimed at strengthening national and regional security for the mitigation of and preparedness against CBRN risks. This includes sharing of good practices and capabilities in terms of expertise, institutional capacity building, training, technical assistance and responding to the needs identified by partner countries. Seven regional centres have been established (Hart 2014).[1] The CoE initiative involves more than 52 countries worldwide and is funded by the EC and implemented in cooperation with the United Nations Interregional Crime and Justice Research Institute (UNICRI). The EEAS is also involved in the follow-up of the initiative (Mignone 2013; Council 2015a).

[1] 1. North Africa: Algiers, 2. Atlantic Façade: Rabat, 3. Middle East: Amman, 4. South East Europe, Southern Caucasus, the Republic of Moldova, Ukraine: Tbilisi, 5. South East Asia: Manila, 6. Sub-Saharan Africa: Nairobi, and 7. Gulf Countries: Abu Dhabi, have started work. Next a Secretariat in Central Asia: Tashkent.

3.3 The Fight Against Terrorism

After the 9/11 terror attacks and the Anthrax letters of 2001 in the US, the EU realized that there were major weaknesses in national and regional emergency preparations and agreed on steps needed in the fight against terrorism (Council 2001a, b). A common EU definition of terrorism was reached in December 2001 that consists of three parts: the context of an action, the aim of the action and the specific acts being committed (Council 2001c). It was further noted that there was a lack of resources in the areas of research and preparedness against CBRN-terrorism. A programme for the protection against biological and chemical terrorism was presented 2001 that was later broadened to cover nuclear and radiological terrorism as well (Council 2001d, 2002).

Aims were to strengthen risk analysis and assessment of CBRN threats, reducing the vulnerabilities of the population, ensure rapid detection and identification of an attack, mitigating the consequences of an attack, and strengthening the scientific basis and cooperate with third countries (Council 2003d, f). The name of the programme was changed in 2004 to the Council solidarity programme to include other types of terrorism besides CBRNE-terrorism (Council 2004c, k).

The terrorist incidents in Madrid (2004), London (2005) and Norway (2011) showed again that terrorism was a major threat to the EU and to international peace and security (Council 2014a). In the event of such terrorist attacks, mutual assistance and collective action are necessary. It is important to control unauthorised access to and prevent trafficking of WMD and related agents and materials (Council 2004d).

Since 2002, military assets and capabilities can be used to assist in protecting civilian populations against the consequences of a terrorist attack, including CBRNE (Presidency Conclusions 2002). In 2004, the military database was made available and capabilities can be used, on a case by case basis, to support civil protection measures against CBRNE terrorism within the EU (Council 2003e, 2006b). In the framework of ECAP (European Capability Action Plan) work has been done on developing CBRNE protection, a CBRN-battalion, a CBRN deployable laboratory, NBC EOD (Explosive Ordinance Disposal) and a CBRN competence centre (Council 2004e). The European Defence Agency (EDA) runs several research projects relevant to CBRNE issues.

A Declaration on Combating Terrorism, including a Declaration on Solidarity against terrorism, as well as the position of Counter-Terrorism Coordinator, and in 2004 an Action Plan on Combating Terrorism (to be updated every 6 months) (Council 2004f, g, j, 2010b, 2015b). In 2005 the European Union Counter-Terrorism Strategy was adopted where all actions are grouped under four headings: prevent, protect, pursue and respond (Council 2005b, 2014a). The Strategy defines the fight against terrorism as being primarily the responsibility of the MS but it demands close cooperation between the Council, the Commission and the European Parliament. In December 2007 the EU decided that effective policies to address CBRNE risks should be developed (Council 2007a). The Council of Europe

Convention on Prevention of Terrorism entered into force in June 2007 (Council of Europe 2005).

In 2008 a revised Framework Decision on combating terrorism was agreed and this was strengthened and updated in 2017. It criminalises: travelling for terrorist purposes, funding such travels, receiving training for terrorist purposes and providing funds to be used to commit terrorist offences (Council 2008c, d; Council/European Parliament 2017). The Hague Programme emphasizes that all the elements of the Action Plan on Combating Terrorism must be implemented (Council 2004h). During the Swedish EU Presidency in 2009 this was further developed into the Stockholm Programme (Council 2009b). It stressed the need for 'a prioritised, relevant and effective European strategy to enhance the protection of EU citizens from incidents involving CBRN materials' (Council 2010c). Several other actions were initiated, like the critical infrastructure protection programme (Council 2004i).

The Internal Security Strategy for the EU 'Towards a European Security Model' was adopted in 2010 (Council 2014b; Venslovaite 2012). Combating cross-border crime and terrorism is a common European responsibility. The strategic objectives set out in the Internal Security Strategy's European Agenda on Security 2010–2014 and for 2015–2020. The Councils Conclusions for 2015–2020 gives priority to terrorism, organized crime, and cybercrime and cybersecurity (Commission 2015a; Council 2015c). The European Council meeting of 15 December 2016, stressed the importance of the political agreement on the Counter-Terrorism Directive, emphasised the need to swiftly adopt the proposals on regulation of firearms and anti-money laundering, as well as the implementation of the new passenger name record (PNR) legislation (European Council 2016). An European Counter-Terrorism Centre (ECTC) within Europol to support MS was opened in 2016 and there is also an internet referral unit in EUROPOL to combat terrorist and violent extremist propaganda. Information sharing on counter terrorism, across European countries as well as through and with Europol, has reached high levels during 2016 (Commission 2015a, 2017a, b; Council of the EU 2016).

Recent terrorist attacks in Europe, have raised the awareness of the vulnerabilities of the open democratic societies in the EU. The attacks did not often fit the profile and modus operandi of previous attacks. The target chosen varied and the attacks were carried out by actors including foreign fighter returnees, home-grown jihadist extremists, and lone actors, with a variety of weapons or explosives. The deadly attacks in Paris, with 17 deaths in January 2015 and another 130 deaths in November 2015, 35 deaths in Brussels in March 2016, six deaths in London in March and another 22 deaths in Manchester in May 2017, have highlighted the reality and extent of the jihadist terrorist threat across Europe (Bigo 2015; RTÉ News 2015). Another trend was using ordinary trucks driven into a crowd to kill people: in Berlin 12 deaths in December 2016, 88 deaths in Nice in July 2016 and 5 deaths in Stockholm in April 2017. According to the EC, recent developments indicate that the threat from CBRNE remains high and is evolving (Commission 2014b). In December 2015 the EC, the European Parliament and the French government warned of a dangerous threat: a terrorist attack with CBRNE agents. The terrorist organizations have financial resources, proven success in recruiting skilled university graduates, and access to CBRNE material, at least in Iraq and Syria, and possibly in Libya (EPRS 2015). They also have an unknown number of sympathisers in Europe. There is an increased probability of a CBRNE attack on European soil. The terrorist groups have allegedly used chemical weapons in Iraq and Syria. One ISIS fighter who was arrested had documents detailing how to weaponise bubonic plague (EurActive 2015; EPRS 2015). Since the recent terror attacks were related to ISIS terrorism, this has been the main EU counter-terrorism focus. Further lone-wolf terrorism is on the increase, facilitated by increased availability of information on the internet and calls upon Muslims in Western countries to commit lone actor attacks in their countries of residence by AlQaeda and more recently ISIS (European Parliament 2017).

3.4 Health Security

MS are responsible for responding to public health and other major emergencies within their borders. The EU's role in the field of public health is limited to complementing the national policies, coordinating MS actions and exchanging data between the EC and the EU MS. The European Centre for Disease Prevention and Control (ECDC) plays a vital role in the epidemiological surveillance, identification, assessment, and communication of current and/or emerging threats to human health and to enhance the EU's ability to respond to public health emergencies (US Library of Congress 2015; Commission 2008a).

Health security has become an area of enhanced EU engagement and the Commission established in 2001 the Health Security Committee (HSC) and established the Health Security Programme (Tegnell et al. 2003). The Health Threat Unit covers CBRN issues and addresses the threat of CBRN agents as a core objective (Grip 2011). The HSC is the key body for risk management of CBRN events in the EU (EU 2010). It is an informal cooperation and coordination body for health–related threats from terrorism or any deliberate release of biological or chemical agents, as well as enhancing preparedness for cross-border threats, in particular for pandemic flu (Commission 2005a). It has strengthened coordination on generic preparedness planning for public health emergencies at the EU level and facilitates an input to the implementation of the International Health Regulations (Commission 2001, 2002, 2005b). The HSC is assisted by a Task Force (Reid 2001) and members exchange information on health related threats, ensure rapid communication in case of major health-related crises, and advise on preparedness and response as well as on coordination of emergency planning at an EU level (Council 2006c, 2007b).

The HEOF (Health Emergency Operational Facilities) helps MS to better handle health problems, epidemics and bioterrorist attacks. It provides the EC with an overview of the health situation and coordinates the MS' command and control activities during emergency operations at Community level (DG SANCO 2007). Measures were adopted in 2013 to enable the EU to respond effectively to serious cross-border health threats and emergencies (Decision 2013a; Commission 2015b).

The EC is also a partner to the Global Health Security Initiative (GHSI) and the Global Health Security Action Group (GHSAG) that focus on coordination and cooperation between different systems for early warning of CBRNE and communicable disease events related to terrorism threats and sharing analytical capacity (GHSAG 2008).

3.5 The EU CBRN Action Plan

The EU adopted a CBRN Action Plan in 2009 based on a report by the CBRN Task Force (Commission 2007, 2009b, c; Council 2008a). Based on an all-hazard approach, the CBRN Action Plan's overall goal is to spread good practices, strengthen interoperability and reduce the threat of, and the potential damage from, CBRN incidents of accidental, natural and malevolent origin, including terrorist acts, and prevent terrorists obtaining CBRN materials. The Action Plan contributes to the implementation of the EU Counter Terrorism Strategy. The Action Plan provides for three main areas of CBRN security work:

- Prevention: ensuring that unauthorized access to CBRN materials of concern is as difficult as possible;
- Detection: having the capability to detect CBRN materials in order to prevent or respond to CBRN incidents;
- Preparedness and response: being able to efficiently respond to incidents involving CBRN materials and recover from them as quickly as possible.

A total of 124 actions were to be implemented by the EU MS and the EU Institutions by the end of 2015. In addition to 25 actions relating to radiological and nuclear security, there are 32 actions covering biological or chemical security. A further 67 actions are horizontal, in the sense that they apply to more than one area. The recommendations included: establish an open-ended list of high risk chemical and biological agents of security concern, monitor industrial use of high risk chemicals, identify facilities possessing listed biological agents, improve security checks of personnel handling chemical or biological agents, and improve security at facilities handling and transporting such agents (European Parliament 2010). In the area of combating illicit trafficking of nuclear and radiological materials, progress has also been made through the implementation of the EU CBRN Action Plan (Commission 2009b) and by outreach activities of the EU. The EU stress tests of all its nuclear power reactors have both a safety and a security track (Commission 2011).

An Action Plan on Enhancing the Security of Explosives (E) was adopted in 2008 (Council 2008e; Commission 2008b). The Plan takes a comprehensive approach to countering threats linked to explosives and precursors to explosives and with an objective to prevent the use of explosive devices by terrorists. The Action Plan envisages the establishment of an Early Warning System concerning explosives, a network of Explosives Ordnance Disposal (EOD) Units active in the civilian context, the establishment of a Standing Committee of Experts concerning precursors and a European Bomb Data System. The need for having a collaborative network of EOD specialists was also identified by EU experts (Commission 2014c). Europol created a European Explosive Ordnance Disposal Network (EEODN) in May 2008 to facilitate cooperation, to share information among EU explosives and CBRN specialists and to organise joint exercises (Commission 2014c).

In order to take the implementation of the EU CBRN Action Plan forward, in early 2010 the Commission established a CBRN Advisory Group and sub-groups (EU 2010). The EU strategy to combat terrorism and the EU CBRN Action Plan are to be implemented by the EC, EEAS, Europol, MS and other stakeholders. DG Home Affairs is responsible for the overall coordination of implementation (Grip 2011).

According to comprehensive progress reports published in 2012 for the CBRN and E Action Plans, implementation was 'relatively uneven' and many of the things that had been done were of a preparatory nature in anticipation of more substantial efforts yet to be undertaken. The reports note the challenge of pursuing all actions in parallel and identified 14 key actions, with a view to ensuring tangible results on these by 2015 (Anthony and Grip 2013; Commission 2012). As the final report of the CBRN and E Actions Plans have been delayed it is not possible to evaluate the results yet.

The CBRN Action Plan foresees that each MS prepares an inventory and assessment of the available medical and technical means to deal with CBRN incidents. The MS will register (on a voluntary basis) self-sufficient, interoperable intervention modules[2] which could be rapidly dispatched within 12 h following a request for assistance. Two of the 17 types of modules that have been defined at EU level are for CBRN detection and sampling, and search and rescue in CBRN conditions. These modules can be supported by assessment and coordination teams which can be sent. This assistance could include expert teams, equipment for the diagnosis, support to health care, vaccines and medicinal products, protective equipment, etc. (Commission 2014c). In 2014, a total of 150 modules and 10 technical and assistance teams have been registered covering different areas (Council 2012a; Dussart 2014).

In 2012 the conference 'A New EU-CBRNE Agenda' in Malmö, Sweden underlined a need to identify areas with insufficient security arrangements, prioritise further common efforts, and study the possibility of creating one comprehensive CBRNE plan (Council 2012b; CBRNE Conference 2012). In May 2014 the EC adopted a new approach to the detection and mitigation of CBRNE risks at EU level with 30 actions as a first step in implementing the new CBRN-E Agenda based on gap analysis. It was also stated that a robust, better designed, and proportionate strategy to anticipate and deter future CBRNE risks at EU level is needed (Commission 2014a, b). The activities focus on improving the existing standards for

[2] The CBRNE modules are characterized by, predefined capabilities, predefined personnel and equipment, rapid deployment, interoperability, self-sufficiency, deployment within and outside the EU, training and exercises financed by EU and resources from one or more Member States.

detection of explosives and CBRN materials (Council 2014a). The European Standardisation Organisation (CEN) has been tasked to develop CBRNE minimum detection and sampling standards (Commission 2014a).

The new CBRN Action Plan builds upon the 2010–2015 CBRN Action Plan. This new Action Plan aims to increase the European cooperation to strengthen CBRN security with a focus on preventing, preparing for, and responding to CBRN threat and terrorism attacks. Many of the proposed actions pursue an all-hazards approach and will also contribute to improving preparedness for any large scale CBRN incidents unconnected to terrorism. To improve coordination at EU level, an EU CBRN security network will pool together all CBRN actors at both strategic (policy-making) and operational levels to overcome the fragmentation of efforts. The network will rely on three pillars: (1) an advisory group bringing together all CBRN security coordinators of the MS, (2) a support network composed of existing CBRN centres across the EU and (3) a CBRN knowledge hub set up in the European Counter-Terrorism Centre (ECTC) in Europol (Commission 2017a).

ARETE 2014 was an exercise of inter-service coordination and enhancement of EU disaster preparedness conducted in Belgium simulating a complex chemical and terrorism situation including hostage-taking (Commission 2015d). Europol conducts several activities that aim at assisting the EU MS in developing their capacity to prevent and respond to CBRNE incidents as well as a CBRNE database of CBRN terrorism-related events and CBRNE products and materials (Grip 2011). In June 2014 Europol and the European Centre for Disease Prevention and Control (ECDC) invited representatives of law enforcement and public health authorities to assess the needs for joint training on the response to the deliberate use of biological agents. In September 2012 the Commission and the US (FBI) organised an EU workshop on Insider Threats in the field of Bio Security (Council 2014a).

A total of 75 CBRNE research projects were funded under the EU Seventh Framework Programme (FP7), followed now by the Horizon 2020 programme but also through the Public Health programme and the Prevention of and Fight against Crime programme with 50 CBRNE projects which provides scientific and technical support to the CBRNE Action Plans covering the whole crisis management from prevention to recovery (Steinhäusler 2012).

3.6 Civil Protection

The Treaty of Lisbon underpins the commitment of the EU to provide assistance, relief, and protection to victims of natural or man-made disasters as well as support and coordinate the civil protection systems of its MS. It further mandates the European institutions to define the necessary measures for such actions to be carried out. The EU Civil Protection Mechanism (EUCPM) was established for cooperation among national civil protection authorities across Europe and was activated for the Ebola outbreak in West Africa (2014), the conflict in Eastern Ukraine (2015) and the European refugee crisis (2015) (Commission 2013; EU Civil Protection Mechanism

2017).[3] EUCPM is the main EU legal instrument for disaster relief interventions inside and outside the EU, facilitating cooperation, applying to both natural and man-made disasters, including acts of terrorism. The EU-response capacity should be developed in accordance with two principles: national responsibility and EU solidarity (Council 2008f). The EUCPM makes available different types of grants for prevention, preparedness and response activities and it was activated 30 times in 2014 (Commission 2015d).

The Emergency Response Coordination Centre (ERCC) is operated by DG ECHO since 2013. It acts as the main 24/7 coordination and support platform for all crises under the EUCPM, the Solidarity Clause and the IPCR. It relies on inputs from the Commission, EU agencies and MS. The ERCC aims to provide a more effective and faster EU response to a disaster (Decision 2013a, b). With increasing and new disaster risks, MS and the Commission need to work together to fully implement and operationalize the 2013 civil protection legislation, including following up on the Sendai Framework for Disaster Risk Reduction 2015–2030 (Commission 2015a). The ERCC replaces and upgrades the functions of the previous Monitoring and Information Centre (MIC). It coordinates the delivery of civil protection assistance to disaster stricken countries such as relief items, expertise, intervention teams and specific equipment. Additionally, the ERCC provides emergency communications and monitoring tools through the Common Emergency Communication and Information System (CECIS) enabling real time exchange of information (Commission Civil Protection 2016).

The EU created the European Emergency Response Capacity (EERC) in 2014, as part of the EUCPM. Any country inside or outside the EU affected by a major disaster can request assistance through the ERCC. The ERCC collects and analyses real-time information on disasters, monitors hazards, prepares plans for the deployment of experts, teams and equipment, and works with participating states to map available assets and coordinates the EU's disaster response efforts by matching offers of assistance to the needs of the disaster-stricken country. The EERC brings together a range of relief teams, experts and equipment, which participating states make available and keep on standby for EU civil protection missions all over the world. This voluntary pool allows for a faster and more effective EU response to disasters and it ensures better planning and coordination of EU operations (Commission Civil Protection 2016). The Commission has also developed risk assessment and mapping guidelines for Disaster Management (Commission 2015a). Directorate General (DG) ECHO has strengthened its role as a hub for crisis management activities in the EU. The EU does not act as a first responder and the primary responsibility for crisis management rests with national governments while the EU's role is mainly supportive.

The EC has also established its internal crisis coordination arrangements called ARGUS with the capability to link all specialized systems for emergencies, and a central crisis centre that brings together all relevant Commission services during an

[3]Currently 2016, 33 states – the EU28 and Montenegro, Norway, Iceland, former Yugoslav Republic of Macedonia and Serbia are participating.

emergency. The system allows each Directorate General in the Commission to inform other Directorates General and services of a beginning or risk of multi-sectoral crisis via an alert exchange. This rapid alert system, under the auspices of the EC, exists to ensure a high degree of coordination between different policy areas of the EC potentially affected in the event of a CBRNE incident (e.g. health and internal security, environment, agriculture, customs, civil protection) (Commission 2005c). ARGUS ensures rapid decision-making and information-sharing among all involved services and rapid alert systems, such as the European Community Urgent Radiological Information Exchange (ECURIE) system for radiological emergencies, the Early Warning and Response System (EWRS) for communicable diseases, the RAS BICHAT,[4] and the Pan European Information System of Explosives Control to Prevent and Fight against Terrorism (Commission 2014c).

As part of the implementation of the CBRN Action Plan, the EC launched a CBRN Resilience Programme in civil protection to support preparedness and enhance effective coordination in response to CBRN incidents. The main objectives of the EU CBRN Resilience Programme are to streamline different strands of work undertaken under the EUCPM. This is achieved by improved linkages between the first responders, including civil protection, health, and law enforcement activities in the field of CBRN, and to tackle identified gaps (Decision 2013b).

The EUCPM disasters caused by acts of terrorism and nuclear or radiological accidents the EUCPM only covers preparedness and response actions within the field of civil protection. The UCPM was activated 30 times in 2014. The EUCPM countries, ECHO and its partners made considerable efforts to respond to the Ebola virus outbreak in western Africa. The ERCC played a pivotal role in coordinating the EU response from the start (Commission 2015d).

The European Emergency Response Capacity (EERC) supports increased preparedness at EU level as regards consequence management of CBRN incidents (Decision 2013b). One is the European Medical Corps (EMC) through which teams and equipment from the EU MS can be rapidly deployed to provide medical assistance and public health expertise in response to emergencies inside and outside the EU. Mobile biosafety laboratories have been developed and deployed during the Ebola outbreak response. The EMC is part of the existing EERC, established under the EUCPM (European Medical Corps. 2016). A Disaster Risk Management Knowledge Centre (DRMKC) offers EU countries technical and scientific advice on their risk assessment methodologies (FloodList 2015).

The Solidarity clause provides the option for the EU and its MS to assist another MS which is the object of a terrorist attack or a victim of a disaster (Fuchs-Drapier 2011). In April 2014 the EC issued the first overview of natural and man-made disasters in the EU covering information for 12 major natural and man-made risks contained in some 17 national risk assessments prepared by Member States (Commission 2014a).

---

[4]RAS BICHAT is the Rapid Alert System used for exchanging information on health threats due to deliberate release of chemical, biological and radio-nuclear agents (notification of confirmed or suspected events).

The EU Integrated Political Crisis Response Capabilities (IPCRC), established in 2013, is a further developed crisis coordination arrangement that allows the EU to take rapid political decisions when facing major crises. An Integrated Situational Awareness and Analysis (ISAA) capability supports the Presidency's and Council's decision-making.[5] IPCRC define rules for interactions between EU institutions and affected MS during a crisis. The EU Integrated crisis management arrangement (EU-ICMA) facilitate practical cooperation between MS for all types of disasters (EU Migration and Home Affairs 2016).

The EU and NATO continue to share information at all levels including inventories listing their respective activities and capabilities for protection of civilian populations against CBRNE terrorist attacks. NATO has enhanced its efforts in the fight against terrorism, preventing proliferation of WMD, and protection against WMD. A task force exists that consists of a CBRN Joint Assessment Team and a CBRN Defence Battalion (NATO 2015).

3.7 Discussion and Conclusions

This overview of the EU as an actor in CBRNE crises is based on proposals and adopted actions as seen in the large number of official documents with relevance for fighting CBRNE terrorism and shows a wide span of actions. This compilation should not be viewed as a complete list, but rather gives the reader a picture how the work in the EU has progressed over time in the various policy areas. The progress varies from area to area and much of the focus has been on agreeing on joint declarations and on judicial matters to enhance coordination and cooperation.

EU crisis management has grown from separate policy sectors, often triggered by unforeseen events, and is a system which consists of separate arrangements and mandates with no or weak linkages between them. The Lisbon Treaty now provides possibilities for improved coordination of emergencies and crises. The EU could appoint a Commissioner for crisis management or contribute to the establishment of an agency tasked to be a promoter of improved crisis management across its policy sectors. The EU could acquire its own civil protection rapid response forces that could be equipped and trained also to support MS in a CBRN incidence if requested to do so.

It is clear that the international non-proliferation measures have been fairly successful, when they are implemented and enforced, in preventing states from engaging in clandestine WMD development. The same cannot be said concerning their effect on preventing terrorists from acquiring CBRNE capabilities as they were not designed for this purpose. There is a trend towards greater international coordination of non-proliferation efforts resulting in greater effectiveness of the measures adopted. The impact of the EU's non-proliferation strategy, however, has been less

[5] The EU integrated political crisis response arrangements, http://www.consilium.europa.eu/en/documents-publications/publications/2014/eu-ipcr/ Assessed 31 January 2016.

than might have been hoped. In terms of key countries and regions of proliferation concern, the EU lacks the leverage to decisively shape the WMD-related decisions of the states concerned (Cottey 2014). A European Agenda on Security for the period 2015–2020 has been proposed by reviewing the ongoing actions and identifying new actions taking into account emerging threats (Commission 2015c; Commission 2016). It is unfortunate that the Global Strategy for its foreign and security policy does not focus more on the risks with proliferation of WMD. This is of concern as for example in the conflict in Syria chemical weapons have been used.

The EU's counter-terrorism agenda since 9/11 shows that most of the policy measures have been put forward as a reaction to major terrorist incidents in New York, Madrid, London, Oslo and Paris. There is a large overlap of counter terrorism policies and other measures adopted by the EU (Venslovaite 2012).[6] It is still not clear which measures the EU will focus on after the most recent terror attacks in 2016–2017. A general problem with the implementation of the EU's antiterrorism policies is that the main powers over these remain in the hands of national governments. It has been argued that the EU counter-terrorism policy has been haphazard and lacked an overall strategy. Others believe that even if measures agreed are unevenly implemented significant progress has been achieved in several areas, such as increased information exchange. (European Parliament 2015). What is most notable about the implementation of these measures is not the varying forms that they have taken at national level, but the fact that MS have frequently been so slow to implement measures agreed, and in a number of cases have not implemented them at all until faced with legal action. Where they have been implemented in whole or part, the Commission's assessments often complain of inconsistency with the EU provisions on which national measures are based (Hayes and Jones 2013). After the terrorist attacks in Paris 2015, Brussels, Nice, and Berlin in 2016, and Paris, London, Manchester and Stockholm so far in 2017.new initiatives in addition to those already agreed are discussed to enhance the fight against terrorism in the EU.

Although the EU adopted the CBRN and E Action Plans with many relevant measures that need to be implemented across Europe, the plans are non-binding and thus cannot be legally enforced. The implementation of the E Action Plan has been fairly good as the plan was more concrete than the one for CBRN. One reason for this difference is that the CBRN Action plan covers many different types of materials and that the plan was not so concrete so as to achieve agreement on it. As a result, the implementation process for the CBRN measures agreed is lengthy or even non-existent in some cases. There should have been a report on the CBRN Action Plan in 2016 but still there is only a draft version that MS have received (Commission 2017b). This is one reason for a somewhat less ambitious 'New Agenda for CBRNE' which is unfortunate.

---

[6] Since '9/11' until 2013, 239 EU counter-terrorism measures have been introduced: 26 action plans and strategy documents, 25 Regulations, 15 Directives, 11 Framework Decisions, 25 Decisions, 1 Joint Action, 3 Common Positions, 4 Resolutions, 111 Council Conclusions, and 8 international agreements as compiled by Ben Hayes and Chris Jones of Statewatch.

References

Alvarez-Verdugo, M. 2006. Mixing tools against proliferation: The EU's strategy for dealing with weapons of mass destruction. *European Foreign Affairs Review* 11 (3): 417–438.

Anthony, I. 2004. *Reducing threats at the source: A European perspective on cooperative threat reduction*, SIPRI Research Report No. 19. Stockholm: Stockholm International Peace Research Institute.

Anthony, I., and Grip l. 2013, June. *Strengthening the European Union's future approach to WMD non-proliferation*, SIPRI Policy Paper No. 37. Stockholm: Stockholm International Peace Research Institute.

Bigo, Didier. 2015. *The EU counter-terrorism policy responses to the attacks in Paris towards an EU security and liberty agenda*, CEPS Paper in Liberty and Security in Europe No. 81/ February. Brussels: Centre for European Policy Studies.

CBRNE Conference. 2012. *A new EU-CBRNE agenda, strategic EU-level, Malmö*. Doc 15365/12. Sweden. 2–3 October.

Commission. 2001. *Programme of cooperation on preparedness and response to biological and chemical agent attacks, health security*. Doc. GFS D (200) GG. European Commission, Luxembourg, 17 December.

———. 2002. Civil protection – Progress made in implementing the programme for preparedness for possible emergencies, Communication from the Commission to the Council and the European Parliament. Doc. 10327/03 COM (2002) 302 final. – Annex I, point 13. European Commission, Brussels, 11 June.

———. 2005a. Communication from the Commission to the Council, the European Parliament, the European Economic and Social Committee and the Committee of the regions, on Pandemic Influenza Preparedness and Response Planning in the European Community. Doc. COM (2005) 607 final. European Commission, Brussels, 28 November.

———. 2005b. Communication from the Commission to the Council, the European Parliament, the European Economic and Social Committee and the Committee of the regions on strengthening coordination on generic preparedness planning for public health emergencies at EU level. Doc. COM (2005) 605 final, European Commission, Brussels, 28 November.

———. 2005c. Communication from the Commission to the European Parliament, the Council, the European Economic and Social Committee and the Committee of the Regions – Commission provisions on ARGUS general rapid alert system. Doc. COM (2005) 662 final. European Commission, Brussels, 23 December.

———. 2007. Green paper on bio-preparedness, Doc. COM (2007) 399 final, European Commission, 11 July.

———. 2008a. Decision setting up an advisory structure of scientific committees and experts in the field of consumer safety, public health and the environment and repealing decision 2004/210/EC, 2008 (L 241) 21, Doc. 2008/721/EC. European Commission.

———. 2008b. Directive setting up pursuant to Council directive 93/15/EEC, a system for the identification and traceability of explosives for civil uses. Doc. 2008/43/EC. European Commission, Brussels, 4 April.

———. 2009a. Report from the Commission to the Council, the European Parliament, the European Economic and Social Committee and the Committee of the Regions, Annual report from the European Commission on the Instrument for Stability. Doc. COM (2009) 341. European Commission, Brussels, 9 July.

———. 2009b. Communication from the Commission to the European Parliament and the Council on Strengthening Chemical, Biological, Radiological and Nuclear Security in the European Union – An EU CBRN Action Plan. Doc. COM (2009) 273 final. European Commission, Brussels, 24 June.

———. 2009c. *Report of the CBRN task force*. European Commission, 13 January.

———. 2011. Communication from the Commission to the European Parliament and the Council on the interim report on the comprehensive risk and safety assessments, "stress tests", of nuclear

power plants in the European Union. Doc. COM (2011) 784 final. European Commission, 24 November.
———. 2012. Progress report on the implementation of the EU CBRN Action Plan. Home Affairs, Doc 10441/12. European Commission, Brussels, 12 May.
———. 2013. Decision on laying down rules for the implementation of decision No 1313/2013/EU of the European Parliament and of the Council on a Union Civil Protection Mechanism and repealing Commission Decisions 2004/277/EC, Euratom and 2007/606/EC. Euratom, European Commission.
———. 2014a. Communication from the Commission to the European Parliament and the Council, The final implementation report of the EU Internal Security Strategy 2010–2014. Doc. COM (2014) 365 final. European Commission, Brussels, 20 June.
———. 2014b. Communication from the Commission to the0 European Parliament, the Council, the European Economic and Social Committee and the Committee of the Regions on a new EU approach to the detection and mitigation of CBRN-E risks. Doc. COM (2014)247 final. European Commission, Brussels, 5 June.
———. 2015a. Communication from the Commission to the Council, the European Parliament, the European Economic and Social Committee and the Committee of the regions, The European Agenda on Security. COM (2015) 185 final. European Commission, Strasbourg, 28 April.
———. 2015b. Report on the implementation of decision No 1082/2013/EU of the European Parliament and of the Council of 22 October 2013 on serious cross-border threats to health and repealing Decision No 2119/98/EC. Doc. COM (2015) 617 final. European Commission, Brussels, 7 December.
———. 2015c. Annex to the Communication from the Commission to the European Parliament, the Council, the European Economic and Social Committee and the Committee of the Regions, Commission Work Programme 2015 A New Start. Doc. COM (2014) 910 final ANNEX 1. European Commission, Strasbourg, 16 December.
———. 2015d. Report from the Commission to the European Parliament and the Council, Annual Report on the European Union's Humanitarian Aid and Civil Protection Policies and their Implementation in 2014. Doc. COM (2015) 406 final. European Commission, Brussels, 21 August.
———. 2016. Communication from the Commission to the European Parliament, the European Council and the Council, Delivering on the European Agenda on Security to fight against terrorism and pave the way towards an effective and genuine Security Union. COM (2016) 230 final. European Commission, Brussels.
———. 2017a. Communication from the Commission to the European Parliament, the Council, the European Economic and Social Committee and the Committee of the Regions, Action Plan to enhance preparedness against chemical, biological, radiological and nuclear security risks, COM(2017) 610 final, European Commission, Brussels, 18 October. https://eur-lex.europa.eu/legal-content/EN/TXT/PDF/?uri=CELEX:52017DC0610&from=EN Assessed 28 Sept 2018.
———. 2017b. *Second progress report on the implementation of the CBRN action plan and the action plan on enhancing the security of explosives*. Commission Staff Working Paper, Brussels 2015 (final version to be published).
———. 2014c. *EU efforts to strengthen nuclear security*, Joint Staff Working Document, European Commission/EU External Action Service, SWD (2014) 107 final, Brussels. 13 March.
Commission Civil Protection. 2016. Humanitarian Aid, European Commission, 9 February. http://ec.europa.eu/echo/what/civil-protection_en. Assessed 28 Sept 2018.
Cottey, Andrew. 2014. The EU's non-proliferation strategy ten years on. *European Foreign Affairs Review, Issue* 1: 45–63.
Council. 2001a. *Common position on combating terrorism*, Doc. 2001/930/CFSP, OJ L334, Council of the European Union, Brussels, 28 December.
———. 2001b. Anti-terrorism road map. Doc. SN4119/01. Council of the European Union, Brussels.

———. 2001c. Proposal for a Council framework decision on combating terrorism. Outcome of Proceedings of the Council. Doc. 14845/1/01. Council of the European Union, Brussels, 7 December.
———. 2001d. Draft EU action plan on the protection of the populations against bacteriological and chemical terrorism. Doc. PROCIV/01/34. Council of the European Union, Brussels, 14 December.
———. 2002. Adoption of the programme to improve cooperation in the European Union for preventing and limiting the consequences of chemical, biological, radiological or nuclear terrorist threats. Doc.14627/02. Council of the European Union, Brussels, 21 November.
———. 2003a. A secure Europe in a better world: European security strategy. Doc. 15895/03. Council of the European Union, Brussels 12 December.
———. 2003b. EU strategy against proliferation of weapons of mass destruction. Doc. 15708/03. Council of the European Union, Brussels, 10 December.
———. 2003c. Fight against the proliferation of weapons of mass destruction – Mainstreaming non-proliferation policies into the EU's wider relations with third countries. Doc. 14997/03. Council of the European Union, Brussels, 19 November.
———. 2003d. Draft guidelines for the implementation of the Council and Commission programme of 20 December 2002 to improve cooperation in the European Union for preventing and limiting the consequences of chemical, biological, radiological and nuclear threats. Doc. 6393/03. Council of the European Union, Brussels, 14 February.
———. 2003e. Adoption of Council conclusions on the database of military assets and capabilities relevant to the protection of civilian populations against the effects of terrorist attacks, including CBRN. Doc. 15564/03. Council of the European Union, Brussels, 3 December.
———. 2003f. First annual presidency report to the Council on the implementation of the joint programme of the Council and the Commission, of 20 December, to improve cooperation in the European Union for preventing and limiting the consequences of chemical, biological, radiological or nuclear terrorist threats. Doc. 16285/03. Council of the European Union, Brussels, 18 December.
———. 2004a. A common position on the universalisation and reinforcement of multilateral agreements in the field of non-proliferation of WMD and their means of delivery. Doc. 2003/805/CFSP. Council of the European Union, Brussels, 17 November.
———. 2004b. Draft public EU statement of support of the Proliferation Security Initiative (PSI). Doc. 9366/1/04. Council of the European Union, Brussels, 10 May.
———. 2004c. Draft second annual Presidency report to the Council on the implementation of the joint Programme of the Council and the Commission, of 20 December 2002, to improve cooperation in the European Union for preventing and limiting the consequences of chemical, biological, radiological or nuclear terrorist threats (2002 CBRN Programme). Council of the European Union, Brussels, Doc. 8988/05, 31 May, 2005.
———. 2004d. Criminal sanctions related to weapons of mass destruction. Doc. 15549/04. Council of the European Union, Brussels, 1 December.
———. 2004e. *Draft Council conclusions on ESDP*. Doc. 14887/2/04 Rev 2. Council of the European Union, Brussels, 22 November.
———. 2004f. Declaration on combating terrorism. Doc. 7906/04. Council of the European Union, Brussels, 29 March.
———. 2004g. PSC contribution to the CFSP/ESDP aspects of the report to the European Council on the implementation of the Declaration on combating terrorism. Doc 10328/04. Council of the European Union, Brussels, 8 June.
———. 2004h. The Hague programme: Strengthening freedom security and justice in the European Union. Brussels: Council of the European Union.
———. 2004i. Critical infrastructure protection in the fight against terrorism. Doc. 13979/04, COM (2004) 702 final. Council of the European Union, Brussels, 26 October.
———. 2004j. EU plan of action on combating terrorism. Doc 10586/04. Council of the European Union, Brussels, 15 June.

———. 2004k. EU solidarity programme on the consequences of terrorist threats and attacks (revised/widened CBRN Programme) – Adoption. Doc. 15480/04. Council of the European Union, Brussels, 1 December.

———. 2005a. Presidency conclusions. Doc. 16238/1/04. Council of the European Union, Brussels, 1 February.

———. 2005b. The European Union counter-terrorism strategy. Doc. 14469/4/05. Rev 4, Council of the European Union, Brussels, 30 November.

———. 2006a. EU strategy against the proliferation of WMD: Monitoring and enhancing consistent implementation. 16694/06. Council of the European Union, Brussels.

———. 2006b. Military support to EU disaster response: – Identification and coordination of available assets and capabilities. Doc. 9462/3/06 REV 3. Council of the European Union, Brussels, 25 October.

———. 2006c. Communication from the Commission to the Council on transitional prolongation and extension of the mandate of the Health Security Committee in view of a future general revision of the structures dealing with health threats at EU level. Council of the European Union, Brussels, Doc. (COM (2006)699 final), 20 November.

———. 2007a. Council conclusions of 6 December 2007 addressing chemical, biological, radiological and nuclear risks and on bio-preparedness. Doc. 16589/07. Council of the European Union, 17 December, p. 6.

———. 2007b. Council conclusions on preparedness for decontamination of casualties following CBRN-incidents, 2897th Justice and Home Affairs Council Meeting. Doc. 10015/07. Council of the European Union, Luxembourg, 12–13 June.

———. 2008a. Report on the implementation of the European security strategy – Providing security in a changing world. Doc. S407/08. Council of the European Union, Brussels.

———. 2008b. New lines for action by the European Union in combating the proliferation of weapons of mass destruction and their delivery systems, annex to council conclusions. Doc. 17172/08. Council of the European Union, Brussels, 17 December.

———. 2008c. Council framework decision amending framework decision 2002/475/JHA on combating terrorism. Doc. 8707/08. Council of the European Union, Brussels, 23 April.

———. 2008d. CTC report – Implementation of the strategy and action plan to combat terrorism. Doc. 9416/08. Council of the European Union, Brussels, 19 May.

———. 2008e. EU action plan on enhancing the security of explosives. Doc. 8109/08. Council of the European Union, Brussels, 4 April.

———. 2008f. Conclusions on reinforcing the Union's disaster response capacity – Towards an integrated approach to managing disasters. Council of the European Union, Luxembourg, 16 June.

———. 2009a. Council Regulation (EC) no. 428/2009 of 5 May setting up a Community regime for the control of exports, transfer, brokering and transit of dual-use items. Council of the European Union, *Official Journal of the European Union*, L 134, 29 May.

———. 2009b. The Stockholm programme: An open and secure Europe serving and protecting the citizens. Doc. 17024/09. Council of the European Union, 2 December.

———. 2010a. Council decision of 26 July 2010 establishing the organisation and functioning of the European, external action service. Doc. 2010/427/EU. *Official Journal of the European Union* L 201, Council of the European Union, Brussels.

———. 2010b. *E*U action plan on combating terrorism. Doc. 15893/10. Council of the European Union, Brussels, 15 November.

———. 2010c. The Stockholm programme – An open and secure Europe serving and protecting citizens 2010–2015. Council of the European Union.

———. 2012a. Annual report on the implementation of the EU counter-terrorism strategy. Doc. 16471/12 ADD 1. Council of the European Union, Brussels, 23 November.

———. 2012b. Draft Council conclusions on the new CBRNE agenda – Adoption. Doc. 16980/12. Council of the European Union, Brussels, 29 November.

———. 2013. Decision in support of World Health Organisation activities in the area of bio-safety and bio-security in the framework of the European Union strategy against the proliferation of weapons of mass destruction. Doc. 2013/668/CFSP. Council of the European Union, Brussels, 18 November.

———. 2014a. Report on the implementation of the EU counter-terrorism strategy. Doc. 15799/14, Council of the European Union, Brussels, 24 November.

———. 2014b. Draft Council conclusions on the development of a renewed European Union internal security strategy. Doc. 14186/1/14, REV 1. Council of the European Union, Brussels, 16 October.

———. 2015a. Six-monthly progress report on the implementation of the EU strategy against the proliferation of weapons of mass destruction (2015/I) as endorsed by the Council (Foreign Affairs) on 20 July 2015, Doc. 10855/15. Council of the European Union, Brussels, 20 July 2015, (2015/C 296/01). *Official Journal of the European Union*, C 296/1-35, 8 September.

———. 2015b. Follow-up to the statement of the members of the European Council of 12 February 2015 on counter-terrorism: Report on implementation of measures. Doc. 9422/1/15, Rev. 1. Council of the European Union, Brussels, 10 June.

———. 2015c. Draft Council conclusions on the renewed European Union internal security strategy 2015–2020, 9798/15, Council of the European Union, Brussels, 10 June.

Council of Europe. 2005. *Convention on the prevention of terrorism*, Warsaw. 16 May. http://conventions.coe.int/Treaty/EN/Treaties/Word/196.doc. Assessed 28 Sept 2018.

Council of the EU. 2016. *I*mplementation of the counter-terrorism agenda set by the European Council. Doc. 14260/16 ADD 1 EXT 1, Council of the European Union, Brussels, 20 December.

Council/European Parliament. 2017. Directive (EU) 2017/541 of the European parliament and of the Council of 15 March 2017, on combating terrorism and replacing Council Framework Decision 2002/475/JHA and amending Council Decision 2005/671/JHA. The European Parliament and the Council of the European Union. Official Journal of the European Union L 88/6-21, 31.3.2017. https://db.eurocrim.org/db/en/doc/2704.pdf Assessed 30 May 2017.

Decision. 2013a. Of the European parliament and of the Council of 22 October 2013 'on serious cross-border threats to health No 1082/2013/EU and repealing Decision No 2119/98/EC. *Official Journal of the European Union*, L 293/1-15, 5 November.

———. 2013b. Of the European parliament and of the Council of 17 December 2013 on a union civil protection mechanism. Doc. 1313/2013/EU. *Official Journal of the European Union,* L 347/924-947, Brussels, 20 December.

DG SANCO. 2007. The Commission health emergency operations facility: For a coordinated management of public health emergency at EU level. Luxembourg: Health & Consumer Protection Directorate-General, European Communities.

Dussart, Yves. 2014. CBRN resilience in civil protection: The EU contribution, Strengthening Science-Policy-Industry links in the CBRN-E sector. DG ECHO, Presentation, Brussels, 30th January 2014.

EPRS. 2015. ISIL/Da'esh and 'non-conventional' weapons of terror', Briefing, European Parliamentary Research Service, Beatrix Immenkamp. No. PE 572.806 EN. December.

EU. 2010. EU capacities to respond to CBRN attacks and CBRN incidents. Submitted by Belgium on behalf of the European Union.

EU CBRN CoE. European Union chemical, biological, radiological and nuclear centres of excellences risk mitigation initiative. http://www.unicri.it/news/article/European_Union_External_Action. Assessed 28 Sept 2018.

EU Civil Protection Mechanism. 2017. *ECHO factsheet – EU civil protection*. European Commission. April.

EU integrated political crisis response arrangements. https://www.consilium.europa.eu/media/29699/web_ipcr.pdf. Assessed 28 Sept 2018.

EU Migration and Home Affairs. 2016. Crisis management. http://ec.europa.eu/dgs/home-affairs/what-we-do/policies/crisis-and-terrorism/crisis-management/index_en.htm. Assessed 28 Sept 201821.

EurActive. 2015. *EU leaders silent on growing risk of an ISIS chemical attack*. 18 December.
European Council. 2016. Conclusions of the European Council meeting of 15 December, EUCO 34/16.
European Medical Corps. 2016. DG ECHO factsheet. European Commission. http://ec.europa.eu/echo/files/aid/countries/factsheets/thematic/European_Medical_Corps_en.pdf. Assessed 28 Sept 2018.
European Parliament. 2010. Report on strengthening chemical, biological, radiological and nuclear security in the European Union – An EU CBRN action plan. Doc. 2010/2114(INI). Brussels, 2 December.
———. 2015. Resolution of 11 February on anti-terrorism measures. Doc.2015/2530(RSP).
———. 2017. The European Union's Policies on Counter-Terrorism, Relevance, Coherence and Effectiveness. Directorate General for Internal Policies, Policy Department C: Citizens' Rights and Constitutional Affairs, Civil Liberties, Justice and Home Affairs Commissioned by the European Parliament's at the request of the LIBE Committee. http://www.europarl.europa.eu/RegData/etudes/STUD/2017/583124/IPOL_STU(2017)583124_EN.pdf. Assessed 28 Sept 2018.
European Union. 2016. Shared vision, common action: A stronger Europe, a global strategy for the European Union's foreign and security policy. Brussels, June. https://eeas.europa.eu/archives/docs/top_stories/pdf/eugs_review_web.pdf. Assessed 28 Sept 2018.
FloodList. 2015. EU launches new disaster risk management knowledge centre, 12 October. http://floodlist.com/europe/eu-launch-new-disaster-risk-management-knowledge-centre Assessed 28 Sept 2018.
Fuchs-Drapier, M. 2011. The European Union's solidarity clause in the event of a terrorist attack: Towards solidarity or maintaining sovereignty? *Journal of Contingencies and Crisis Management*, December, 19(4)184–197.
GHSAG, W.G. 2008. On risk management and communication. Meeting on Early warning for CBRN threats, Luxembourg – 26–28 February.
Grip, Lena. 2011. Mapping the European Union's institutional actors related to WMD non-proliferation. *Non-Proliferation Papers*, No. 1, May. Stockholm International Peace Research Institute.
Hart, John. 2014. Chapter 8, Part III: Biological weapons and arms control and disarmament. In *SIPRI Yearbook* 2014, p. 413. Armaments, Disarmament and International Security, Stockholm International Peace Research Institute, Oxford University Press.
Hayes, Ben, and Chris Jones. 2013. *Report on the transposition of EU counter-terrorism measures*. SECILE November 26, http://www.statewatch.org/news/2013/nov/secile-report-on-the-transposition-of-eu-counter-terrorism-measures.pdf Assessed 28 Sept 2018.
Lundin, Lars-Erik. 2017. The European Union and weapons of mass destruction: A follow-on to the global strategy? EU Non-Proliferation Consortium. *Non-proliferation Paper*. no. 58. Stockholm International Peace Research Institute. May.
Mignone, A. 2013. The European Union's chemical, biological, radiological and nuclear centres of excellence initiative EU Non-proliferation consortium. *Non-Proliferation Papers*, No. 28 June.
NATO. 2015. Combined joint chemical, biological, radiological and nuclear defence task force, The Alliance's multinational CBRN defence capability, 6 Aug. http://www.nato.int/cps/it/natolive/topics_49156.htm. Assessed 28 Sept 2018.
Presidency Conclusions. 2002. Contribution of CFSP, including ESDP in the fight against terrorism. Seville Council 21–22 June 2002. Doc. SN 200/1/02 REV 1.
Reid, F. 2001. Health ministers wish to strengthen the network approach in response to bioterrorism threat. *Eurosurveillance Weekly* 5: 011129.
Roffey, R. 2011. *United Nations actions against international terrorism, can it become more effective?* Swedish Defence Research Agency, Stockholm, FOI Report R 3176 SE, April.
RTÉ News. 2015. Paris attacks death toll rises to 130, 20 November. http://www.rte.ie/news/2015/1120/747897-paris/. Assessed 28 Sept 20182.

Steinhäusler, F. 2012. Gap analysis of EU counterterrorism research initiatives. Erice International Seminars on Planetary Emergencies, 45th Session: The Role of Science in the Third Millennium, Erice. Italy, 19–25 August. http://www.federationofscientists.org/PlanetaryEmergencies/Seminars/45th/ www.federationofscientists.org/.../Steinhausler%20. Assessed 28 Sept 2018.

Tegnell, A. et al. 2003. The European Commission's task force on bioterrorism. *Emerging Infectious Diseases*, Vol. 9, No. 10 October.

UNGA. 2007. UN general assembly, negotiations on comprehensive anti-terrorism convention at critical stage says chairman as committee concludes eleventh session. Department of Public Information, New York, L/3114, 15 February. http://www.un.org/News/Press/docs/2007/l3114.doc.htm Assessed 28 Sept 2018.

UNSC. 2004. UN Security Council Resolution 1540. 28 April.

———. 2008. Committee, S/2008/493 – Report of the committee established pursuant to resolution 1540 (2004), UN Security Council, 8 July. https://undocs.org/S/2008/493 Assessed 28 Sept 2018.

US Library of Congress. 2015. European Union: Legal responses to health emergencies. https://www.loc.gov/law/help/health-emergencies/europeanunion.php Assessed 28 Sept 2018.

Van Ham, P. 2011. *The European Union's strategy on weapons of mass destruction, from ambition to disappointment*. Clingendael Paper No. 6, Netherlands Institute of International Relations 'Clingendael', p. 13.

Venslovaite, Egle. 2012. *The European Union counter terrorism agenda: The effects of the attacks in Norway on the 22nd of July, 2011*. Department of Public Administration Leiden, The Netherlands, 17-May.

Chapter 4
Building Ethics into CBRNE Security

Andrew P. Rebera

Abstract This chapter explores the most effective way to build ethics into CBRNE security. The most obvious approach may appear to encompass generating a list of activities involved in CBRNE security and to develop ethical guidance for each item. But this approach encounters two significant difficulties. First, every decision in CBRNE security has an ethical side and so the list is likely to be extremely long. Second, in this domain ethical issues are often insufficiently distinguished from societal and legal issues: this is liable to obscure important nuances. All this makes the main practical goal of CBRNE ethics – to build ethics into all aspects of CBRNE resilience – a challenging one. In this chapter I argue that the need to support responders in on-the-spot ethical decision-making, amid so much pressure and so many stressors, suggests that organisations should build ethics into CBRNE security by developing an ethos of key ethical values and principles. Staff should be trained in how to operationalise the ethos when making ethical decisions. Finally, I argue that the complexity of CBRNE security – its multi-agency, multi-disciplinary, inter-regional, and international character – requires us to focus on how organisations with divergent ethical positions, ethoses, and cultural backgrounds can cooperate effectively. Such cooperation should provide clarity and reassurance so that CBRNE professionals can make on-the-spot ethical decisions relatively free from doubt as to the legitimacy of their decisions in uncertain cultural contexts.

Keywords CBRN ethics · CBRNE ethics · Ethical decision-making · Ethos · Global bioethics

A. P. Rebera (✉)
AND Consulting Group, Brussels, Belgium
e-mail: rebera@and-cg.com

D. P. O'Mathúna, I. de Miguel Beriain (eds.), *Ethics and Law for Chemical, Biological, Radiological, Nuclear & Explosive Crises*, The International Library of Ethics, Law and Technology 20, https://doi.org/10.1007/978-3-030-11977-5_4

4.1 Introduction

Political, scholarly, and public interest in CBRNE security – security issues raised by the possibility of incidents (whether accidentally or intentionally caused) involving chemical, biological, radiological, nuclear or explosive (CBRNE) materials or agents – has increased rapidly in recent years. In the European Union (EU) for instance, 2009 saw the adoption of the 'CBRN Action Plan' (Council of the European Union 2009). The Action Plan is a policy package aiming to reduce the likelihood and impact of potential CBRNE incidents by enhancing existing national (Member State) capacities through cooperation, coordination, and funding. Building on a review of the Action Plan (European Commission 2012a), as well as of a related 'Action Plan on Enhancing the Security of Explosives' (European Commission 2007, 2012b), the European Commission adopted a new CBRNE Agenda in 2014 (European Commission 2014).

Given this interest, it is surprising that the many ethical questions and challenges raised by preparation for, response to, and recovery from CBRNE incidents (henceforth I refer to these questions and challenges as 'CBRNE ethics') have received relatively little direct attention. Reference to ethics in EU CBRNE policy is, generally, only indirect. Often it is couched in terms of 'negative social impacts' (European Commission 2009: 7), or, as in the Action Plan, in terms of international law and human rights (Council of the European Union 2009: 7). In academic and practitioner-focused literature also, direct references to CBRNE ethics are few.[1] The latter is partly explained by the overlap between CBRNE ethics and other fields of applied ethics. To take just a few examples, ethical dimensions of decisions concerning the prioritisation of vaccines or prophylactic drugs (Moodley et al. 2013), disaster triage (Barilan et al. 2014; Ten Have 2014; Wagner and Dahnke 2015), or the obligations and rights of healthcare professionals (Eckenwiler 2004; Grimaldi 2007), have been (and continue to be) addressed in medical ethics, bioethics, and disaster bioethics. Disaster bioethics – the investigation of ethical questions raised by the occurrence of disasters and by the needs of those affected by them (O'Mathúna et al. 2014) – as well as the more general fields of disaster, crisis and emergency management, cover many of the operational issues raised by CBRNE security (Carter et al. 2013; Davis and McHenry 2005).

This chapter starts with two assumptions. The first is that the main practical goal of CBRNE ethics is to build ethics into all aspects of CBRNE security (those who find this controversial may substitute 'a' for 'the': '*a* main practical goal …'). The second assumption is that the goal can be achieved by developing and implementing guidelines, codes of conduct, and training to support responders and anyone else

[1] In an (extremely) informal survey of Google Scholar, keyword searches (within article titles) returned a combined total of 939 articles for the terms 'CBRN', 'CBRN-E', and 'CBRNE'. The addition of the keywords 'ethics' or 'ethical' reduced total returns to just 2 articles. (Keyword 'CBRN' returned 693 articles; 'CBRN, ethics' returned 0; 'CBRN, ethical', 2. 'CBRN-E' returned 5 articles; no returns for 'CBRN-E, ethics' or 'CBRN-E, ethical'. 'CBRNE' returned 241 articles; no returns for 'CBRNE, ethics' or 'CBRNE, ethical'. Search conducted 15 May 2016.)

involved in CBRNE security in handling ethically sensitive situations and decisions. Neither assumption will be defended here. Of the first, let me merely state that – in contrast to the second – it seems self-evident. The second is relied upon here only because guidelines, codes of conduct, and training have, as a matter of fact, been the main tools for organisations and professional bodies (of all kinds) to address ethical issues in their domains. I have no doubt that such tools are an important – essential even – means of (further) embedding ethical behaviours in any organisation; but it seems likely, in the case of CBRNE security (as well as of disaster management generally), that additional methods could support and expedite the process. The deployment of technology – decision support tools, for instance – could promote ethical behaviour if it has been through some process of 'ethical design' aimed at mitigating certain potential ethical problems. As a very simple example, consider the potential to minimise responders' burdens to handle personal data collected in the field responsibly by minimising the amount of such data collected in the first place by a tool. On the other hand, it is also important to focus on developing the ability of responders and others involved in CBRNE security to recognise and respond to ethically-charged circumstances. Guidelines, codes of conduct, and training are all valuable here; but the way in which they are used is critical. Ultimately, ethics must be embedded in CBRNE security through the sensitivity of those involved to the ethical aspects of the domain: it will not be so embedded merely through the existence of guidelines setting out processes for handling a given situation.

How, then, should we set about developing guidelines, codes of conduct, and training (henceforth I will often just use 'guidelines' for short) for building ethics into the various activities involved in CBRNE security? The obvious approach is – to put it somewhat crudely – to generate a list of activities involved in CBRNE security (at each stage of the process from planning and preparation to response and recovery) and provide, for each activity, a (set of) guideline(s). Those guidelines should be informed by, *inter alia*, findings already put forward in the fields mentioned above (disaster bioethics and the like).

Although I see no reason to suppose the obvious approach misguided in any major sense, certain difficulties should be acknowledged. One is that *everything* in this domain is ethical. Or to put it more precisely: every decision taken in the context of CBRNE security has an ethical side to it.

(1) A question, activity, problem, decision …, (etc.) has an ethical side to it just in case it has to do with the wellbeing of one or more moral-agents.
(2) All decisions taken in the context of CBRNE security have to do with people's well-being (in particular their security in the face of the possibility their being caught up, in one way or another, in a CBRNE incident). *So:*
(3) All decisions taken in the context of CBRNE security have an ethical side to them.

Both premises in this argument require clarification. But while such clarifications as emerge may temper the force of the conclusion, they are unlikely to rebut its general thrust. To illustrate this, consider some set of decisions having to do with the culling of livestock in the context of an ongoing incident. Considering premise (1) we may ask whether livestock (or non-human animals generally) are genuine moral agents? Perhaps (a) livestock are moral agents; perhaps (b) they are not; and perhaps (c) there is a position between these two extremes – say, that they have interests and rights but are not full moral agents. Without settling this, let it be noted that: if (a), the conclusion stands; if (c), then rephrasing (1) in terms of rights and interests will suffice; and if (b) then the argument will stand if, in addition to the wellbeing of the livestock, we consider the wellbeing of any (uncontroversial) moral agents connected to the animals by ownership, affection, or other morally significant ties. In addition to the question of moral agency, clarifications concerning the nature of 'wellbeing', as well as of the conditions on a decision's 'having to do with' a moral agent's wellbeing are raised by premise (1). In relation to the first of these, it is spectacularly implausible that any adequate account of wellbeing would undermine its conceptual link with security. In relation to the second, I contend that a decision (or its possible consequences) need only be very indirectly connected to a person's wellbeing to morally compel the decision-maker to take *some* account of that person's interests (other things being equal, the less direct the connection the less weight those interests carry). Admittedly this raises a number of further questions. However, since this is not the place to address those questions, and since most decisions in CBRNE security (or their consequences) are relatively directly related to some *identifiable* person's or community's interests or wellbeing (i.e. the extreme cases that might be thought to undermine the contention are not in play), let us accept, for present purposes, the general thrust of the overall argument. Note that acceptance of a watered-down version – e.g. that *most* CBRNE security decisions have an ethical side – will suffice to ground the claims made in this chapter.

The list of activities involved in CBRNE security is very long: CBRNE ethics raises a lot of issues. A second difficulty, mentioned in passing above, is that, in the policy domain especially, ethical issues are often insufficiently well distinguished from questions of societal impact or from legal issues concerning fundamental rights. This is liable to obscure important nuances. Standard means of measuring societal impact (quasi-economic and quantitative techniques, for example) are not always appropriate for measuring ethical impact. And similarly, while legal arguments about fundamental or human rights are relevant to certain philosophical arguments about ethics, the law and philosophy remain distinct disciplines with their own concepts, methods, and conventions: caution is required whenever carrying over an argument from one domain to the other. Thus CBRNE ethics is liable to be somewhat sprawling and messy: *sprawling* because every action or decision has an ethical aspect of some colour and intensity; *messy* because ethical aspects of policy and planning often need disentangling from societal and legal ones. All of this makes the main practical goal of CBRNE ethics – i.e. of building ethics into all aspects of CBRNE resilience – a challenging one.

In this chapter I address the question of how best to build ethics into CBRNE security. In Sect. 4.2 I argue that the possibility of CBRNE professionals being faced with genuine ethical dilemmas imposes two responsibilities upon organisations involved in CBRNE security: firstly, to explicitly develop and communicate to their staff an 'ethos' of values and principles to guide action and decision-making; secondly, to train their staff in how to 'act out' that ethos – how to operationalise the values and principles – when making on-the-spot ethical decisions. In Sect. 4.3, reflecting on difficulties in resolving the role of culture and cultural diversity in bioethics, I argue that instead of developing a 'common global ethos' for CBRNE security, efforts should focus on how organisations with divergent ethical positions and cultural backgrounds can cooperate effectively in spite of their differences.

4.2 Operationalising Values in On-the-Spot Ethical Decision-Making

Difficult ethical decisions abound in disaster preparation and response (O'Mathúna et al. 2014: 5). CBRNE incidents raise the genuine possibility of ethical dilemmas: i.e. lose-lose situations for responders in which decisions must be taken amid extreme time-pressure, information gaps, and other stressors to decision-making (Karadag and Hakan 2012; Rebera and Rafalowski 2014). Ethical challenges that could be expected in general disaster management or emergency medicine are likely to be exacerbated by the presence of hazardous materials in CBRNE security. For example, administering drugs (Castle et al. 2010), conducting field triage (Ramesh and Kumar 2010), and gathering patient consent (Rebera and Rafalowski 2014) may all be more difficult in CBRNE incidents due to the use of PPE (personal protective equipment, such as hazmat suits). Exposure to hazardous materials (e.g. from off-gassing or treatment of infected victims) is a risk that responders must take seriously. Moreover, tension between competing professional and personal responsibilities is also possible in very serious or large-scale CBRNE incidents. To take just one example, the duty of care that healthcare professionals bear to their patients cannot be simply assumed to outweigh personal interest in their own wellbeing, nor the responsibilities owed to loved ones (Sokol 2006). (It should be noted also that the management (tracking, controlling of access to, etc.) of hazardous materials raises ethical issues even in the absence of an incident).

The goal here is not to create a comprehensive list of possible dilemmas in which on-the-spot ethical decisions are required, but to think about how such situations can best be handled. I have elsewhere urged the importance of the following factors in on-the-spot ethical decision-making in CBRNE contexts (Rebera and Rafalowski 2014).

(i) The responder must decide, with urgency, between competing plausible courses of action, each of which carries some serious negative consequence.
(ii) Contextually-specific factors – which could render a candidate course of action unviable, but which cannot be predicted in advance – are likely to be relevant to the responder's decision.
(iii) To help guide and focus the decision-making process, the responder should have a relatively simple overriding goal to aim for (e.g. *to save life*).

(iv) Notwithstanding point iii, it is of the very nature of the dilemmas under discussion that other values will compete with any overriding goal the responder aims for.

Whereas many operational decisions and procedures can be largely guided by standard operating procedures (SOPs), the above points – ii and iv especially – show that *standard* procedures are unlikely to adequately support responders in *non-standard* situations, faced with the kinds of dilemmas under discussion. Standard codes of conduct for disaster response, such as that produced by the International Federation of Red Cross and Red Crescent Societies (IFRC 1996), are 'highly aspirational and would require further exploration to permit practical application' (O'Mathúna et al. 2014: 5). They simply cannot make reference to the contextually-specific factors of a given scenario; nor do they reflect the large body of research in disaster management suggesting the importance of creativity and improvisation in adjusting to the chaotic and unpredictable circumstances of large-scale crises (Mendonca and Fiedrich 2006; Webb 2004).

In line with a general obligation to promote ethically acceptable treatment of persons affected by CBRNE incidents, but also in line with an obligation to minimise the psychological impact on responders (Schwartz et al. 2014), organisations need to provide better support to their staff in the ethically complex situations they can expect to face. To this end, Rebera and Rafalowski (2014) propose a 'modified consequentialist approach' to on-the-spot ethical decision-making. On this view, a central value or principle – saving lives, for example – forms the basis of a goal-oriented heuristic. Additional core rights and values are factored-in as 'side-constraints' (Nozick 1974; cf. Kinslaw et al. 2009), i.e. minimum standards beyond which any violation is unacceptable.

The modified consequentialist approach is by no means perfect; but it does offer a relatively flexible basic framework (since nothing dictates that the central value be *saving lives* – though it seems a reasonable choice – and the values acting as side-constraints can be any that an organisation chooses). The point here is not to defend this particular approach *per se*, but simply to insist upon *some form* of robust support to responders in resolving ethical dilemmas under conditions (i)–(iv). The basis for any such support is, I suggest, an institutional ethos of values and principles that shape and guide actions and decisions within an organisation.

A clear statement of an organisation's ethos should set the tone for all activities undertaken, and for all decisions taken, by its representatives. It should set out the most important values and principles that the organisation wishes to build into its work. An organisation's existing written ethos (if they have one), mission statement, the IFRC code of conduct (1996), or standard professional values – such as the core bioethical principles of autonomy, beneficence, non-maleficence, and justice (Beauchamp and Childress 1994) – may be suitable starting points. However, in order to effectively support a decision-making heuristic, at least two further points should be noted. First, an ethos must recognise that priorities may change in the event of, or during, an incident (ACP 2012: 37). Second, and more fundamentally, significant and ongoing effort is required to ensure that the values given by an ethos can be readily operationalised, i.e. translated into actions and decisions in the field. It does not suffice to simply *identify* the core values; and while providing better *definitions* of values and principles will improve an ethos statement or code of conduct (Borovecki et al. 2015), this is not enough either. Practical guidance on operationalising values and principles is required. A high-level ethos or code of conduct must be accompanied by guidance on how the core values and principles are, in a practical sense, to be respected in relation to key tasks, and, moreover, how they are to be respected in novel or unexpected situations.

'Key tasks' are activities in which ethical decisions are required (or in which failure to act in a certain way could have negative ethical consequences). Such activities for emergency responders in CBRNE security might include the conducting of disaster triage, decontamination, evacuations, dealing with the public, effective communication while in PPE, and so on. General guidelines for such tasks should be developed by an organisation, incorporating the particular values contained in their ethos. These guidelines should set out the overriding goal of the activity (e.g. to save lives) as well as any side constraints (e.g. respect for autonomy, and others). Critically, the guidelines must speak to such difficulties of interpretation as can be foreseen. For instance, the overriding goal may be to save lives. But when difficult choices have to be made, which lives (if any) will be prioritised? Regarding side constraints, which values will be so adopted? And what minimum standards will be set? At what point, for instance, does the overriding of a person's autonomy constitute an unacceptable interference with their rights? In many cases, it will be possible to set standards that can be immediately applied by a responder in a wide range of scenarios. To take a couple of fairly simple examples, guidelines on when patient consent can be assumed (when the patient is unconscious, say) can be specified in advance; and for decontamination (of people), acceptable interferences with the right to privacy can be set in advance (e.g. to determine the use of water-curtains in public view).

Though standards of this nature can be set in advance, the possibility of contextually-specific defeating factors – which can give rise to novel or unexpected situations – can never be ruled out. Only the responders or other professionals on scene can judge the presence and nature of contextual factors. In order to do so, they must be well-versed in the values and principles of the ethos of their organisation, and attuned to signals indicating that those values are threatened or under pressure. They should also be as attuned as possible to signals indicating that the values, principles, and interests of the communities they are serving are threatened or under pressure – especially when operating in unfamiliar socio-cultural settings (this kind of issue is addressed in Sect. 4.3). Such awareness requires sensitivity and a kind of openness. Once recognised, the capacity to deal with such situations requires creativity and innovation (Mendonca and Fiedrich 2006; Webb 2004), as well as awareness of the impact of stress, cognitive bias and moral framing on judgement and decision-making (Greene et al. 2008; Starcke et al. 2012; Sunstein 2007).

The suggestion then is this. Organisations active in CBRNE security should develop a sophisticated ethos, which sets out core goals as well as the values and principles that shape the way in which its representatives pursue them. This ethos should be operationalised, both through the production of ethics guidelines for handling key tasks, and through training and the development of heuristic tools to support responders in value-oriented on-the-spot ethical decision-making. To be sure, none of this is at all simple. Implementing the suggestion requires of an organisation a significant commitment of time, effort, and money – resources which are all at a premium. But if we take seriously the ethical challenges of CBRNE security (and of disaster management in general) – in particular the urgency, complexity, and presence of ethically-salient, contextually-specific factors in many possible scenarios, to say nothing of the potential psychological impact on responders – then the commitment seems unavoidable. Beyond the (already difficult) questions concerning key tasks, the nature of the kinds of ethical challenges foreseeable in large-scale CBRNE incidents *forces* the question of values upon any organisation that sends responders out into those situations: the development and operationalisation of a sophisticated institutional ethos seems a basic requirement.

4.3 Complexity and Cooperation: Do We Need a 'Common Global Ethos'?

CBRNE security is a multi-agency, multi-disciplinary pursuit, involving healthcare professionals, public health organisations, emergency responders and senior officers, political decision-makers of various levels and seniority, specialist agencies and research laboratories, the military, and many others. A CBRNE incident may constitute a *disaster*, defined by the United Nations International Strategy for Disaster Reduction as 'a serious disruption of the functioning of a community or a society […] which exceeds the ability of the affected community or society to cope using its own resources' (UNISDR 2009: 9). Disasters require inter-region or perhaps international support; but even independently of disasters, CBRNE incidents may be international, requiring cooperation across borders, and may trigger solidarity commitments between states or regions. For example, the Lisbon Treaty commits the EU and its Member States, through a solidarity clause, to support any one of their number which is victim to a terrorist attack or natural or man-made disaster (Gestri 2012: 109ff). There is potential for considerable organisational fragmentation, operational incompatibility and confusion. Interests, spheres of influence, professional practices and SOPs – as well as codes of conduct and ethics – are liable to clash.

Additional complexity derives from the forceful impact of potential CBRNE incidents on the general public. CBRNE security has, in this sense, a relatively high profile, with the low-likelihood-massive-impact nature of CBRNE incidents (at least as they are popularly imagined) generating a sort of lurid fascination (Chandler 2009). Moreover, the association of CBRNE security with terrorism has consequences for the public's evaluation of the risks involved (Brooke Rogers et al. 2007). CBRNE security thus plays directly into existing political and societal tensions concerning various kinds of extremism, community cohesion, supposed tension between security and (other) fundamental rights, inequality, and so forth. The potentially widespread impact of CBRNE incidents (their capacity to entirely disrupt a society) means that many communities and groups have important insights into and perspectives on CBRNE security. Businesses seek continuity in the face of economic disruption; minority communities may seek reassurances about cohesion and non-discrimination; and civil society and advocacy groups may lobby in support of various civil liberties, fundamental rights, and other interests. CBRNE security is thus politically sensitive in ways that, say, natural disasters are not (though this is not to deny the socio-political side of natural disasters: exposure and vulnerability to natural disasters are known to track demographic and socio-economic factors such as gender, ethnicity, wealth, education, literacy, lack of political voice, and so on (Cutter et al. 2003; Brooks et al. 2005)). Yet for policymakers to consult every interest group will be difficult at the best of times and impossible in the midst of an acute incident. Moreover, the manner in which any such consultation process is conducted is likely to reflect existing power structures and injustices (democratic deficits, failures of transparency, and so on).

CBRNE security is, then, by its very nature, a highly complex field requiring inter-organisational, inter-regional, and international cooperation. Effective coordination, whether through bilateral agreements or through global or international bodies such as international humanitarian organisations (e.g. the International Red Cross and Red Crescent Movement, *Médecins Sans Frontières*, etc.), the United Nations, or the World Health Organisation (WHO), is essential. If inter-organisational, inter-regional, and international cooperation are required in order to ensure effective coordination of policy and operational aspects of CBRNE security, and if every (or *most*, or even *some*) policy and operational decisions has (have) an ethical side, then some form of inter-organisational, inter-regional, and international cooperation on ethics is needed.

How is this global cooperation on CBRNE ethics to be brought about? In the context of this chapter, the question comes down to this: is it feasible to develop a 'common global ethos for CBRNE security', i.e. a statement of values and principles, which shapes and guides actions and decisions within an organisation, and which is endorsed by organisations across diverse geographical, political, and cultural contexts? A common global ethos would be the basis for ethical decision-making by CBRNE professionals, from whatever organisation, cooperating to prepare for, respond to, and recover from CBRNE incidents.

The prospects for such an initiative do not, initially at least, seem good. The role of cultural diversity in bioethics is a matter of some dispute. There is widespread agreement that cultural issues are in some way relevant to decision-making in bioethics. Respect for cultural diversity is, for example, firmly established as a principle of disaster bioethics and humanitarianism. For example, the Code of Conduct for the International Red Cross and Red Crescent Movement and NGOs in Disaster Relief includes a commitment to 'respect the culture, structures and customs of the communities and countries we are working in' (IFRC 1996); and the WHO 'global guidance' on ethical considerations in influenza pandemic preparedness plans explicitly states that 'specific decisions will depend on local circumstances and cultural values' and that, as such, its recommendations should be adapted 'to the regional and country-level context' (WHO 2007: 2).

Yet there is disagreement as to the exact nature of the role of cultural diversity in bioethics. According to Bracanovic, the 'growing insistence on the importance of culture and cultural differences' (2011: 229) in bioethics has two main roots. The first is the view that Western approaches to bioethics – understood to be based on a conception of universal ethical values applicable in all situations – are too simplistic to be adequate to the complex cultural situations in which they are applied. (In fact, two distinct points are conflated here: that a universalistic 'one-size-fits-all' approach is overly simplistic (Irvine et al. 2002); and that the Western approach is blind to its own culturally-based assumptions (Koenig and Marshall 2003; Brody 1997).) The second is that a more inclusive approach, which takes other (i.e. non-Western) cultural traditions more seriously, will open bioethics to alternative and sometimes better understandings of ethical problems.

However, 'universalists' such as Bracanovic play down the normative and ethical significance of cultural factors, focussing on them as of more empirical and anthropological importance. Ten Have and Gordijn (2011: 2), for instance, acknowledge the value of understanding how ethical problems are conceptualised and tackled in cultures other than one's own, but classify this as a form of 'exploration and clarification' which should serve as 'just [...] the starting-point for what seems to be essential in genuine bioethics: normative analysis'. Bracanovic notes that, in an effort to avoid moral relativism, proponents of 'culturally oriented bioethics' are prepared to set aside some beliefs, no matter how deeply entrenched in a culture, if they are morally indefensible. This, he argues, condemns the cultural approach to vacuity: 'To insist on respect for cultural diversity in bioethics, but only insofar as cultural influences remain "benign" [...], is simply too trivial to be "critically relevant" for bioethics' (Bracanovic 2011: 235).

On the other hand, according to Chattopadhyay and De Vries (2013), the universalists fail to reconcile the important role of socio-cultural context with the impulse to posit culture-free bioethical principles. Reaching this impasse, Chattopadhyay and De Vries (2013) claim, the universalists retain their universalism and problematize respect for cultural diversity; whereas it is, in their view, better – indeed it is an *ethical imperative* – to retain full respect for cultural diversity and to abandon universalism.

The dispute between universalists and culturally-oriented bioethicists is a tangled one. I raise it here not to enter into it, but in order to highlight some of the conceptual challenges facing any attempt to develop the kind of 'global cooperation' or 'common global ethos' mooted in this section. Three possibilities for surmounting the challenge of generating cooperation in the face cultural-diversity suggest themselves. The first, and easiest to dispense with, is to hope for an agreement on the role of cultural factors in bioethics. This is unlikely to happen any time soon.

The second is to bring together stakeholders in bioethics and CBRNE security from around the world to agree 'framework principles' or 'shared values'. These need not be derived from (purportedly) universal principles, but may result from dialogue and inter-cultural exchange. Such initiatives are certainly valuable exercises; yet absent common agreement on the role of cultural diversity in bioethical argument, a framework of principles developed in this way would, in all likelihood, be either so vaguely stated as to be difficult to apply, or it would appeal to so-called 'shared values' which were, in reality, subject to divergent interpretation in different cultural settings. As Qiu (2004: 1) puts it: 'although these shared values are useful in practice' – which is to say that, as decoded in some particular cultural setting, they might be of use in solving ethical problems – 'their content is too poor to constitute an overarching universal ethics or global bioethics'. To put it another way, insofar as such shared values are useful, they are not genuinely shared (I have my useful value and you have yours); but insofar as they are genuinely shared, they are not useful.

The third and most viable possibility for generating cooperation in the face cultural-diversity, is to have each actor in CBRNE security unilaterally settle their own approach to CBRNE ethics (their ethos), and then, rather than seeking consensus on a single shared ethos, seek consensus on how organisations can cooperate in spite of differences of ethics and culture. Of course, to the extent that agreements on shared approaches to ethical problems can be found, they should be vigorously pursued. But where this is not possible, the focus should switch to developing inter-organisational, inter-regional, and international agreement on accountability for decision-making in cross-cultural settings. That is to say, organisations need to come together to establish agreed approaches to resolving situations in which the values and principles which guide their decisions clash. For example, if a responder makes a decision which is in conformity with the ethos of their organisation, but which is not in conformity with either the established norms of the socio-cultural context in which they are operating, or with the ethos of one or more other organisations active in the response, there is potential for controversy, disagreement, mistrust, or worse. The proposal is that a consensus should be sought (in advance) on how such situations are to be resolved. In particular, given the difficulties that responders face in making on-the-spot ethical decisions (as discussed in Sect. 4.2), it is important to ensure that their ability to bring to bear a clear, goal-oriented heuristic is not compromised by confusion or doubt as to the legitimacy of their decisions in unfamiliar or uncertain cultural contexts. Responders should be able to go about their work confident that the principles they have been trained to follow can be applied: that is, that their decisions or actions will not cause harm or undue offence; and that to the extent that their decisions or actions do, inadvertently, cause harm or offence, they will not be held personally responsible (assuming that they acted in conformity with their training and established best-practice in their organisation).

A natural consequence of this approach is that local organisations would be strongly encouraged to take a lead in activities wherever possible (since their values are most likely to match those of the affected communities). This should go some way towards reducing concerns about 'moral imperialism' (Chattopadhyay and De Vries 2013; Qiu 2004). Organisations would be acting in accordance with their own values (their own ethos), but would not be forcing their values upon anyone else. The consensus would be an agreement on *how to cooperate* – not an agreement on a single set of behaviours, and not an agreement on a universal ethical view. No particular position on the status of universalism, relativism, or anything else in bioethics is presupposed. In case of a very profound disagreement on some issue, two organisations might be forced to agree not to collaborate (in some activities or at all). Naturally that would be a shame, and to the extent that collaboration was essential to the wellbeing of the affected community, it is to be hoped that ongoing discussion would result in some form of collaboration. But even if collaboration is not agreed upon, at least the position of each organisation is out in the open and therefore subject to criticism and popular or political pressure.

4.4 Conclusion

This chapter has been concerned with the question of how best to build ethics into CBRNE security. Starting from the most obvious approach – namely to generate a list of activities involved in CBRNE security and then develop ethical guidance for each item – we have been driven, by the nature of CBRNE incidents, to more demanding positions. The need to support responders in on-the-spot ethical decision-making, amid extreme time-pressure, information gaps, and other stressors such as the presence of hazardous materials and the use of PPE, led us to propose that any organisation involved in CBRNE security should develop an 'ethos' of key ethical values and principles. Staff should be trained in how to operationalise the values and principles embedded in the ethos when making on-the-spot ethical decisions. The complexity of CBRNE security – its multi-agency, multi-disciplinary, inter-regional, and international character – then led us to address the question of whether to move from an organisational to an international level ethos: a 'common global ethos for CBRNE security', endorsed by organisations across diverse cultural contexts. However, intractable disagreement on the role of cultural diversity in bioethics led us to, in effect, bracket that question, and focus instead on how organisations with divergent ethical positions and cultural backgrounds can cooperate effectively in spite of their differences. This form of cooperation is intended to provide clarity and reassurance so that responders and other CBRNE professionals can make on-the-spot ethical decisions free from doubt as to the legitimacy of their decisions in uncertain cultural contexts.

It could hardly be claimed that any of the tasks to which this chapter has attempted to draw attention are easy. And yet if we deem CBRNE ethics important, these tasks cannot be avoided: the need for them follows directly from the nature of CBRNE incidents. That said, what is presented in this chapter is merely a sketch; an outline of a roadmap perhaps. Further work is required to develop, extend, and challenge the 'modified consequentialist approach' to on-the-spot ethical decision-making referred to in Sect. 4.2 (Rebera and Rafalowski 2014); and in conjunction, a deeper account of the idea of an 'organisational ethos' is required. The development of a framework for cooperation in the face of divergent ethical and cultural positions is as much a political challenge as it is a conceptual issue in disaster bioethics; and much work is required to clarify both what such a framework might look like as well as how it can be agreed upon.

It was said in the introduction that the main practical goal of CBRNE ethics – i.e. of building ethics into all aspects of CBRNE resilience – is a challenging one. This chapter cannot claim to have made that challenge any less daunting. If it has any value, it consists in its having begun to make the nature of the challenge clearer.

References

ACP (American College of Physicians). 2012. American College of Physicians ethics manual, 6th edition. *Annals of Internal Medicine* 156: 73–104.

Barilan, Y. Michael, Margherita Brusa, and Pinchas Halperin. 2014. Triage in disaster medicine: Ethical strategies in various scenarios. In *Disaster bioethics: Normative issues when nothing is normal*, ed. Dónal P. O'Mathúna et al., 49–64. Dordrecht: Springer.

Beauchamp, T.L., and J.F. Childress. 1994. *Principles of biomedical ethics*, 4th ed. Oxford: Oxford University Press.

Borovecki, Ana, Jónína Einarsdóttir, Dónal O'Mathúna, Paulina Pospieszna, Orly Maya Stern, and Natália Oliva Teles. 2015. 20 years of the ICRC Code of Conduct for Disaster Relief: what do we need to improve? *The Lancet* 385: 1391.

Bracanovic, Tomislav. 2011. Respect for cultural diversity in bioethics. Empirical, conceptual and normative constraints. *Medicine, Health Care, and Philosophy* 14: 229–236.

Brody, H. 1997. Medical bioethics and cultural diversity. *Indian Journal of Pediatrics* 64: 277–284.

Brooke Rogers, M., Richard Amlôt, G. James Rubin, Simon Wessely, and Kristian Krieger. 2007. Mediating the social and psychological impacts of terrorist attacks: The role of risk perception and risk communication. *International Review of Psychiatry* 9 (3): 279–288.

Brooks, Nick W., Neil Adger, and P. Mick Kelly. 2005. The determinants of vulnerability and adaptive capacity at the national level and the implications for adaptation. *Global Environmental Change* 15 (2): 151–163.

Carter, Holly, John Drury, G. James Rubin, Richard Williams, and Richard Amlôt. 2013. The effect of communication during mass decontamination. *Disaster Prevention and Management: An International Journal* 22 (2): 132–147.

Castle, Nick, James Bowen, and Neil Spencer. 2010. Does wearing CBRN-PPE adversely affect the ability for clinicians to accurately, safely, and speedily draw up drugs? *Clinical Toxology* 48 (6): 522–527.

Chandler, J. 2009. Privacy versus national security: clarifying the trade-off. In *Lessons from the identity trail: Anonymity, privacy and identity in a networked society*, ed. I. Kerr et al., 121–138. Oxford: Oxford University Press.

Chattopadhyay, Subrata, and Raymond De Vries. 2013. Respect for cultural diversity in bioethics is an ethical imperative. *Medicine, Health Care and Philosophy* 16 (4): 639–645.

Council of the European Union. 2009. Council conclusions on strengthening chemical, biological, radiological and nuclear (CBRN) security in the European Union – An EU CBRN Action Plan. http://register.consilium.europa.eu/doc/srv?l=EN&f=ST%2015505%202009%20REV%201. Accessed 27 Sept 2018.

Cutter, Susan L., Bryan J. Boruff, and W. Lynn Shirley. 2003. Social vulnerability to environmental hazards. *Social Science Quarterly* 84 (2): 242–261.

Davis, Stephen C., and Kathryn E. McHenry. 2005. A retrospective analysis of mass casualty presentation resulting from the release of toxic chemicals. *International Journal of Emergency Management* 2 (3): 231–238.

Eckenwiler, Lisa A. 2004. Ethical issues in emergency preparedness and response for health professionals. *AMA Journal of Ethics, Virtual Mentor* 6 (5). http://journalofethics.ama-assn.org/2004/05/msoc2-0405.html. Accessed 27 Sept 2018.

European Commission. 2007. Communication from the Commission to the European Parliament and the Council on enhancing the security of explosives. http://eur-lex.europa.eu/legal-content/EN/TXT/PDF/?uri=CELEX:52007DC0651&from=EN. Accessed 27 Sept 2018.

———. 2009. Commission staff working document accompanying the communication from the Commission to the European Parliament and the Council on strengthening chemical, biological, radiological, and nuclear security in the European Union – An EU CBRN Action Plan. http://www.eumonitor.eu/9353000/1/j4nvgs5kjg27kof_j9vvik7m1c3gyxp/vi7jgtc84gyb/f=/11480_09_add_2.pdf. Accessed 27 Sept 2018.

———. 2012a. Progress report on the implementation of the EU CBRN Action Plan, May 2012. http://ec.europa.eu/dgs/home-affairs/what-we-do/policies/crisis-and-terrorism/securing-dangerous-material/docs/eu_cbrn_action_plan_progress_report_en.pdf. Accessed 27 Sept 2018.

———. 2012b. Progress report on the implementation of the EU Action Plan on enhancing the security of explosives (public version). https://ec.europa.eu/home-affairs/sites/homeaffairs/files/what-we-do/policies/crisis-and-terrorism/explosives/docs/progress_report_on_explosives_security_2012_en.pdf. Accessed 27 Sept 2018.

———. 2014. Communication from the Commission to the European Parliament, the Council, the European Economic and Social Committee and the Committee of the Regions on a new approach to the detection and mitigation of CBRN-E risks. http://ec.europa.eu/dgs/home-affairs/what-we-do/policies/crisis-and-terrorism/explosives/docs/20140505_detection_and_mitigation_of_cbrn-e_risks_at_eu_level_en.pdf. Accessed 27 Sept 2018.

Gestri, Marco. 2012. EU disaster response law: Principles and instruments. In *International disaster response law*, ed. A. de Guttry et al., 105–128. The Hague: TMC Asser Press.

Greene, J.D., S.A. Morelli, K. Lowenberg, L.E. Nystrom, and J.D. Cohen. 2008. Cognitive load selectively interferes with utilitarian moral judgment. *Cognition* 107 (3): 1144–1154.

Grimaldi, Mary Elizabeth. 2007. Ethical decisions in times of disaster: Choices healthcare workers must make. *Journal of Trauma Nursing* 14 (3): 163–164.

IFRC (International Federation of Red Cross and Red Crescent Societies). 1996. The Code of Conduct for the International Red Cross and Red Crescent Movement and NGOs in Disaster Relief. International Review of the Red Cross 310. https://www.icrc.org/eng/resources/documents/article/other/code-of-conduct-290296.htm. Accessed 27 Sept 2018.

Irvine, R., J. McPhee, and I.H. Kerridge. 2002. The challenge of cultural and ethical pluralism to medical practice. *Medical Journal of Australia* 176: 174–175.

Karadag, Ozge C., and A. Kerim Hakan. 2012. Ethical dilemmas in disaster medicine. *Iranian Red Crescent Medical Journal* 14 (10): 602–612.

Kinslaw, K., D.H. Barrett, and R.J. Levine. 2009. Ethical guidelines in pandemic influenza: Recommendations of the Ethics Subcommittee of the Advisory Committee of the Director, Centers for Disease Control and Prevention. *Disaster Medicine Public Health Preparation* 3: 185–192.

Koenig, B.A., and P.A. Marshall. 2003. Anthropology and bioethics. In *Encyclopedia of bioethics*, ed. S.G. Post, 3rd ed., 215–225. New York: Thomson Gale, Macmillan Reference.

Mendonca, D., and F. Fiedrich. 2006. Training for improvisation in emergency management: Opportunities and limits for information technology. *International Journal of Emergency Management* 3 (4): 348–363.

Moodley, K., K. Hardie, M.J. Selgelid, R.J. Waldman, P. Strebel, H. Rees, and D.N. Durrheim. 2013. Ethical considerations for vaccination programmes in acute humanitarian emergencies. *Bulletin of the World Health Organization* 91 (4): 290–297.

Nozick, R. 1974. *Anarchy, state and utopia*. New York: Basic Books.

O'Mathúna, Dónal P., Bert Gordijn, and Mike Clarke, eds. 2014. *Disaster bioethics: Normative issues when nothing is normal*. Dordrecht: Springer.
Qiu, Ren-Zong. 2004. Bioethics and Asian culture – A quest for moral diversity. In *Bioethics: Asian perspectives: A quest for moral diversity*, ed. Ren-Zong Qiu, 1–9. Dordrecht: Kluwer.
Ramesh, Aruna C., and S. Kumar. 2010. Triage, monitoring, and treatment of mass casualty events involving chemical, biological, radiological, or nuclear agents. *Journal of Pharmacy and Bioallied Sciences* 2 (3): 239–247.
Rebera, Andrew P., and Chaim Rafalowski. 2014. On-the-spot ethical decision-making in CBRN (chemical, biological, radiological or nuclear event) response. *Science and Engineering Ethics* 20 (3): 735–752.
Schwartz, Lisa, Matthew Hunt, Lynda Redwood-Campbell, and Sonya de Laat. 2014. Ethics and emergency disaster response. Normative approaches and training needs for humanitarian healthcare providers. In *Disaster bioethics: Normative issues when nothing is normal*, ed. Dónal P. O'Mathúna et al., 33–48. Dordrecht: Springer.
Sokol, D.K. 2006. Virulent epidemics and scope of healthcare workers' duty of care. *Emerging Infectious Diseases* 12 (8): 1238–1241.
Starcke, K., A.C. Ludwig, and M. Brand. 2012. Anticipatory stress interferes with utilitarian moral judgment. *Judgment and Decision making* 7 (1): 61–68.
Sunstein, C.R. 2007. Moral heuristics and risk. In *Risk: Philosophical perspectives*, ed. T. Lewens, 156–170. London: Routledge.
Ten Have, Henk. 2014. Macro-triage in disaster planning. In *Disaster bioethics: Normative issues when nothing is normal*, ed. Dónal P. O'Mathúna et al., 13–32. Dordrecht: Springer.
Ten Have, Henk, and Bert Gordijn. 2011. Travelling bioethics. *Medicine, Health Care, and Philosophy* 14 (1): 1–3.
UNISDR (United Nations International Strategy for Disaster Reduction). 2009. *UNISDR terminology on disaster risk reduction*. Geneva: UNISDR.
Wagner, Jacqueline M., and Michael D. Dahnke. 2015. Nursing ethics and disaster triage: applying utilitarian ethical theory. *Journal of Emergency Nursing* 41 (4): 300–306.
Webb, G.R. 2004. Role improvisation during crisis situations. *International Journal of Emergency Management* 2 (1–2): 47–61.
WHO (World Health Organisation). 2007. *Ethical considerations in developing a public health response to pandemic influenza*. Geneva: WHO Press.

Chapter 5
Ethical Frameworks for CBRNE Crises: Toward Shared Concepts and Their Practical Application

Irene Anne Jillson

Abstract For more than a half-century in the field of bioethics, ethical principles generally considered applicable across religions and national boundaries and embedded in international guidelines included autonomy, beneficence, respect for human dignity, and justice. For more than a decade, there has been a flourishing of philosophical, social, and other discourse regarding health ethics, much of which has benefited from a rich diversity of scholars and experts who have explored health ethics from a wide range of perspectives, including bioethicists from the Democratic Republic of the Congo and South Africa, philosophers from Cameroon and India, public health physicians from Egypt, and biomedical researchers from Argentina and Venezuela. Given that we live in an increasingly global system, consideration of these new concepts of and frameworks for bioethics is essential for all those engaged in preventing and/or addressing the aftermath of chemical, biological, radiological, nuclear or explosive (CBRNE) crises. This chapter presents several of these frameworks, as well as an alternative global ethical framework relevant for CBRNE. An example of the practical application of this framework – to the use of "big data" to prevent and address consequences of CBRNE threats – is also presented.

Keywords Bioethics framework · Cultural values · Bioethics principles · CBRNE ethics · Disaster ethics

5.1 Our Shared Heritage of Bioethics

Notwithstanding the popular perception that bioethics or health ethics was a twentieth Century development based largely on European philosophy, concern for and social responses to ethical issues have been extant for millennia. Documentation of ethical principles dates to the Code of Hammurabi—circa 2500 BC—which

I. A. Jillson (✉)
Georgetown University, Washington, DC, USA
e-mail: iaj@georgetown.edu

D. P. O'Mathúna, I. de Miguel Beriain (eds.), *Ethics and Law for Chemical, Biological, Radiological, Nuclear & Explosive Crises*, The International Library of Ethics, Law and Technology 20, https://doi.org/10.1007/978-3-030-11977-5_5

included provisions concerning the importance of ethical considerations to clinical practice and recognized the physician's dual responsibility to the patient and to society. The Codex stipulated that if patients were not satisfied, they could seek justice from a legal system, and punishment if a physician erred "through omission or commission." It was designed "… to cause justice to prevail in the land, to destroy the wicked and the evil, to prevent the strong from oppressing the weak … to further the welfare of the people … to [ensure that] the strong might not oppress the weak" (Spiegel 1997, p. 7). Among the earliest bioethics principles are those traced to Pharaonic Egypt during the period 4000–3500 BC. The Goddess MAAT represented the concept of truth, justice, righteousness, balance, and order, calling for individual responsibility for the community, including in health and in care for the vulnerable (Karenga 2003).

Hindu philosophy holds sacred the transcendent character of human life, expressed through the principles of sanctity and quality of life, and the duty to preserve and guard individual and communal health and Buddhism has a focus on social embeddedness of individuals (Cheng-Tek Tai 2008). In Judaism, *Tikun olam* is the Judaic imperative to repair the world, reflecting the divine values of justice (*tzedek*), compassion (*hesed*), and peace (*shalom*) (Goldsand et al. 2001). Christianity values charity toward individuals and members of society generally and the sanctity of human life (Pellegrino 2005). Islam takes into account the physical, mental and spiritual needs of individuals, as do other faiths, and also holds that, given the dignity of human life, it must be protected (Shomali 2008).

Among these spiritual belief systems, the shared ethical values include: benevolence; charity/mercy; care for the whole person (physical, mental, social, and spiritual); justice, including rights of individual and distributive justice; social good as essential (care for society); and rights and dignity of the individual (sanctity of human life). Knowing that there are common values across spiritual beliefs is critical for an understanding of the developing conceptual frameworks of health ethics, in particular given the often associated – indeed, deeply embedded – link between and among culture, social context, and spiritual beliefs of communities. Although early Western bioethics tended to be *perceived* as secular/non-religious, it was intertwined with the culture and context of those who have been speakers for the field – publishing articles and other documents, speaking to a wide range of audiences, and creating or influencing public policy related to bioethics. Given the predominance of Judeo-Christianity in the U.S., in which the academic field of bioethics was primarily formed, the field has been embedded primarily, although not exclusively, in these spiritual beliefs and in Western philosophy. Indeed, three of the original institutional centers of the discipline are the Institute of Religion at the Texas Medical Center, founded in the early 1950s; the Hastings Center, founded in 1969 by Daniel Callahan, a well-known Catholic philosopher and theologian; and the Kennedy Institute of Ethics at Georgetown University, founded in 1971 (Drane 2002). The more secular roots have been traced to societal reactions to the revelations of human experimentation, in particular the revelations from the Nuremberg trials and the revelations by Beecher of experiments with retarded children in New York, as well as to participation in democratic protests, which expanded significantly in the

1960s. In the U.S., this civic activism focused in anti-Vietnam war protests, civil rights and women's rights (Stevens 2014). In European and other countries, the civic activism also addressed concerns particular to each country.

At the institutional and national levels, we are increasingly aware of the need to address the diverse cultures and contexts in our healthcare systems and in the increasingly global research enterprise. This is particularly important given that health ethics has been applied to health issues as diverse as clinical practice; protection of human and animal subjects of biomedical, behavioural and social research; resource allocation; environmental ethics; and preventing and/or addressing the aftermath of chemical, biological, radiological, nuclear or explosive (CBRNE) crises. As collaborative engagement in these issues expands – in many ways narrowing the geographic and socio-political distance between regions and countries – recognition of the role of diverse spiritual beliefs and of bioethics is vital.

Notably, at the field's nascence, there were essentially two worldviews. The first was that of Potter and colleagues, who "urged using the term to signify ethical analysis of health, well-being, and global survival understood as a webbed function of human beings interconnected with their environments" (Stevens 2014, p. 2). Rather, the more limited view of the Kennedy Institute of Bioethics took root, focusing on medical settings rather than on broader issues and contexts, and expressed as the widely-known and referenced principlism approach (Stevens 2014). The Principles of Biomedical Ethics (principlism) posits four basic principles: autonomy (respect for individual's rights), beneficence (the ethical obligation to maximize benefits, non-maleficence (minimizing harm to individuals), and justice (treating individuals fairly, equitably) (Beauchamp and Childress 1979). These principles were considered by the authors of principlism to be universal, embedded in both ethical principles and medical codes (McCarthy 2003). That is, they were believed to transcend geographic, cultural, economic, and political boundaries. Indeed, they have served as a basis for clinical ethics; the conduct of research involving human subjects at the national, regional, and international levels; and the training of clinicians and researchers. However, in the ensuing decades, in particular as the application of bioethics has expanded beyond its use in clinical services and research, the conceptualization and interpretation of these domains has evolved. This is true, for example, in terms of justice, which is increasingly considered with regard to equitable allocation of resources ranging from financial expenditures for health services to human organs for transplantation, and consideration of environmental ethics, including public and private sector responsibility for both protecting the environment and remediation of environmental degradation.

The International Ethical Guidelines for Biomedical Research Involving Human Subjects, a seminal document in global bioethics, was issued in 1993 by the Council for International Organizations of Medical Sciences (CIOMS 1993). CIOMS initially set out, in cooperation with WHO, to prepare guidelines to indicate how the ethical principles set forth in the Declaration of Helsinki of the World Medical Association, could be effectively applied. The organization did so by paying particular attention to low-resource settings, given their socio-economic circumstances, laws and regulations, and executive and administrative arrangements. Since then revised editions of the CIOMS ethical guidelines have been published, most recently in 2016 (CIOMS 2016a). CIOMS also published ethical guidelines related to specific areas of medical science, for example, vaccine safety communication (2018) and evidence synthesis and meta-analysis for drug safety (2016b). Interestingly, in the new 2016 version of the ethical Guidelines, CIOMS provides answers to a number of pressing issues in research ethics. The Council does so by stressing the need for research having scientific and social value, by providing special guidelines for health-related research in low-resource settings, by detailing the provisions for involving vulnerable groups, and by including major sections of discourse from theological voices (the early contributions from Catholic, Jewish and other sources such as Paul Ramsay) and utilitarian secular voices such as Peter Singer.

However, there have been increasing calls from diverse individuals and institutions for a fuller, more global understanding of the essential morality, spirituality and social justice foundations for bioethics; indeed, beyond *bio*ethics, to consider a broad range of cultures, contexts and worldviews, (e.g., O'Neill 2002; Rennie and Mupenda 2008; Farmer and Campos 2004). While it has long been recognized that bioethics is not religion ipso facto, the dominance of Western bioethics for decades after the 1960s increasingly raised concerns that there are primary religious or cultural influences. This is important generally given the complexity of our global health system and the other sectors with which it is inextricably related – national and global economies, socio-political structures, and conflicts.

5.2 Towards Global Health Ethics

Over the past decade, there has been a flourishing of philosophical, social, and other discourse regarding health ethics; much of this discourse has been based in new concepts of and approaches to health research, development, and public health. Importantly, the discourse has benefited from a rich diversity of scholars and others who have explored health ethics from a wide range of perspectives. New concepts of health ethics principles and frameworks have been proposed by bioethicists, philosophers, public health physicians and biomedical researchers from many countries, including for example, Cameroon (Tangwa 1999), Uganda and South Africa (Benatar et al. 2003), Egypt (El Setouhy et al. 2004) and Japan (Sakamoto 1999).

Among the conceptual principles and frameworks that have been posited—and that have been the basis for discourse regarding health ethics—are the following:

- The Human Development Approach to International Research (London 2005), which makes explicit the linkages between medical research and the social determinants of health and global justice.
- The Fair Benefits Approach (El Setouhy et al. 2004), which concerns distributive justice in the context of medical research.
- Ethics from Below (Farmer and Campos 2004), which addresses health ethics more broadly and suggests contextualizing ethical dilemmas broadly in the

social sciences, rather than focusing only on moral philosophy, as well as in the culture and context of a wide range of settings, including low-income countries and areas/populations within countries, and advocates for the systematic participation of the community in exploration of ethical issues and decision-making.
- A Participatory Framework for Health Ethics (Singer and Benatar 2001), which also addresses the importance of decision-makers, researchers, and communities addressing bioethics issues from their own perspective, engaging in partnerships and strategic alliances to do so as appropriate and necessary.
- The African Meta-physical Worldview, which is influenced by the ecological, cultural and biological diversity that has shaped African social systems and indigenous ethical, religious and metaphysical views that prevail in the region. It also emphasizes the interdependence among and peaceful coexistence with earth, plants, animals and human beings (Tangwa 1999)—now commonly referred to as *one health*.
- The Asia-Pacific School of Ethics, which focuses on "holistic harmony" in contrast to Western "dualistic individualism" – on social "well-orderedness" rather than on individual interests, including assignment of social roles to individuals (Sakamoto 1999, p. 194).

Rather than focusing somewhat narrowly on clinical services or medical research, many of the new approaches consider public health, access to individual health care and to related services, on the broad socioeconomic context of health and on environmental ethics – and the intersections among these. Moreover, as they reflect a wide diversity of cultures and religions, they are not embedded only in Western philosophical theories or Judeo-Christian principles. Although such approaches served for more than half a century to endeavour to ensure protection of human subjects of research and, to a lesser extent, equitable access to health care, consideration of religious, cultural, and other values and norms beyond those prominent in the West has enriched the more recent discourse and enabled a broader understanding of both health ethics generally and its implications for CBRNE.

Awareness of these new concepts of health ethics is essential for all those engaged in addressing ethical issues related to CBRNE, which is more aligned with these broader concepts. Applying the Global Health Ethics Framework of Benatar, Daar, and Singer, for example, requires considering not only philosophical paradigms, but also allowing "the extension of human rights beyond civil and political rights to include social, economic and cultural rights and their close integration with reciprocal responsibilities" (Benatar et al. 2003, p. 121). Thus, the individual's rights in the face of prevention of CBRNE and responding to CBRNE incidents would be considered in the context of interpersonal ethics (including freedom of choice), public health ethics (including equity and access to emergency response services), and environmental rights, which pertain to all members of society. However, these individual rights are also balanced with collective rights of his or her society, the global community, and individual and collective responsibilities.

Those engaged in this rich discourse have not abandoned the essential principles of bioethics that have been a foundation of medical and health practice and research for more than half a century, but rather have expanded our understanding of these principles. For example, in the case of the principle of autonomy – with a focus on *individualism* – it is recognized that ethical policies and practices must consider the need to balance individual rights with social good. The Asian school of bioethics has contributed importantly to this discourse, noting the importance of "holistic harmony" in contrast to Western "dualistic individualism" (Sakamoto 1999, p. 194). Similarly, the principle of justice includes a focus on protection of human rights generally, not just with regard to protection of research subjects, but also with regard to their access to efficacious pharmaceuticals that result from their participation in clinical trials, and on allocation of resources for health services and in the context of disaster relief. The concept of beneficence has broadened to recognize the interdependence and peaceful coexistence between the earth, plants, animals and human beings—the intersection of human, animal and environmental ethics.

Given these considerations, how should we define global health ethics? The United Nations Educational, Scientific and Cultural Organization (UNESCO) definition of bioethics is:

> …analysis of ethical issues raised by life-sciences, the application of technology and, medicine and health policies. It encompasses all fields of scientific development which affect human beings socially, judicially and environmentally. (Feinholz 2015)

The 2015 report *The UNESCO Universal Declaration on Bioethics and Human Rights: Background, principles and application* includes an extensive analysis of the divergent concepts and definitions of bioethics and the implications of these definitions for the development of both the 1993 Declaration and the programs and projects that have ensued. Kirby, for example, notes that, "The deletion of 'social' sciences appears to be deliberate and designed to limit the scope of the ethical issues addressed by the Declaration" (Kirby 2015, p. 75). Nonetheless, UNESCO has played a significant role in the development and promulgation of bioethics, in particular with regard to helping to ensure both a broader global reach and strengthening capacity of low-resource countries to implement bioethics policies and practices through its Global Ethics Observatory and other programs and resources. More than a decade ago, Andorno noted both the important contributions of UNESCO to the development of international conventions regarding bioethics, including the Universal Declaration on the Human Genome and Human Rights, and the significance of the use of a human rights framework as the basis for its original and subsequent declarations and activities (Andorno 2007).

Taking into account the recent rich discourse regarding bioethics in the global context, a broad definition of global health ethics might be:

> A set of widely shared foundational values with respect to health, combining genuine respect for the dignity of all people with a recognition that health and human development are inextricably linked with human rights, economic opportunities, good governance, peace and development. (Jillson 2011)

5.3 Application of a Global Bioethics Framework to CBRNE in Practice

Why is it important now to consider Alternative Concepts of Bioethics with respect to CBRNE? We live in an increasingly global "system" with multiple, in some cases conflicting international, multilateral, and national and even intra-national policies, regulations and programs intended to address prevention of and responses to CBRNE threats. For example, the intersection between and among data systems – "big" data designed to share among governments agencies at multiple levels information regarding possible perpetrators of CBRNE threats – poses ethical concerns if not conundrums. These include ensuring balancing the protection of privacy (autonomy, individual rights) with the responsibility for protecting members of society. The rapid development and diffusion of technology designed to prevent and respond to CBRNE and other threats also poses ethical concerns – an example is the development of drones and autonomous weapon systems (AWS). Caton raised a number of ethical issues regarding AWS in particular, noting the concern with respect to relinquishing moral judgement to either the machines or their creators, and the ability of society to be engaged in decisions regarding their use. He also suggests that ethical and legal considerations of such technologies should be considered at all stages of development (Caton 2015).

A significant distributive justice issue in preventing and addressing the consequences of CBRNE is the allocation of resources from public funds for: local planning and security measures (e.g., equitable distribution by local jurisdiction); for the purchase and distribution of individual protective equipment (e.g., equitable distribution by type of first responder); for provision of health services for those harmed by CBRNE events (e.g., equitable distribution of supplies and services); and for environmental remediation necessary following a CBRNE event (e.g., equitable allocation of resources by community. Throughout, a broad ethical consideration is the degree to which and ways in which the society generally and the community(ies) that are at-risk or that have been affected are engaged in transparent discourse regarding preparedness planning for CBRNE and immediate response to events. To ensure equitable participation, such engagement must be based, insofar as possible, on relevant, valid data and information; recognizing and addressing the complexity of setting priorities and allocating resources for preventive action, intervention, and post-crisis response; and the ethics of trade-offs – balancing societal and individual rights and the roles and responsibilities of emergency responders.

A 2014 report by the EDEN Project described practical ethical issues related, for example, to the roles and management of volunteers and healthcare workers (EDEN Consortium 2014). With respect to the former, the report notes both that management of volunteers is related to the ethical principle of beneficence and to the ultimate objective of disaster response to reduce harm and that there are important issues of informed consent that must be addressed. With respect to healthcare workers, an oft-mentioned ethical conundrum is that of the duty to care – What risks should they be required or expected to make? How can or should they choose between duty to society and duty to family? The report includes a discussion of "approaches to on the spot ethical decision-making for first responders," with a summary of principles that underlie such approaches. These focus on consequentialism and the alternative proposed by the authors, which includes six principles: restriction of individual liberty, proportionality, reciprocity, clarity, transparency and trust, solidarity, and respect for human dignity, non-discrimination and equity. This issue is also addressed by O'Mathúna in this volume.

The European Commission's PBIED Ethics Guideline Document included an ethical framework for consideration. It was based on European normative ethical theories (consequentialism, deontology and virtue ethics) and included a discussion of the Universal Declaration of Human Rights and International Humanitarian Law (Stanciugelu 2015). Neither of these documents has a reference to ethical principles embedded in philosophy or spirituality other than that found in Judeo-Christian faith and culture. This is unfortunate, in particular given the recognition by the EDEN report in particular of the importance of the need to consider carefully ethical considerations in urgent "on the spot" decision-making in CBRNE response.

The following conceptual framework is presented as a point of departure for discussion of the application of global health ethics to CBRNE (Fig. 5.1).

To ensure that the framework is global in the broadest possible sense – that is, that it takes into account differences in individual values or beliefs and normative societal ethics—each of the four domains would need to be considered from diverse

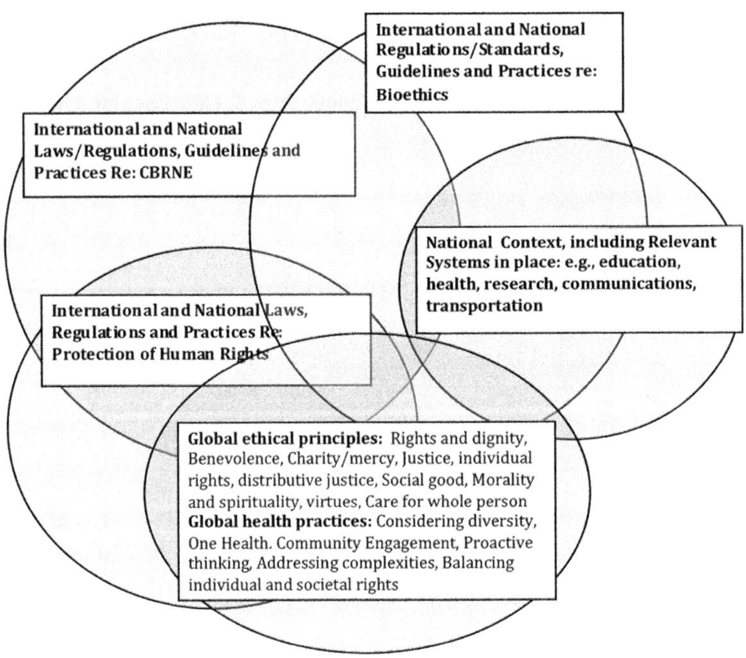

Fig. 5.1 Alternative global bioethics for CBRNE framework Jillson, IA, 2015–iaj@georgetown.edu

perspectives. For example, *global health ethical principles and practices* would consider not only those embedded in Western philosophy (and Judeo-Christian spirituality) but in that which derives from the fullest possible range of philosophy and religion, as well as secular practices. The ethical principles listed in this domain are those that are common, if differently applied in practice within and across religions. There is a sufficiently rich body of literature in bioethics to take these differences into account as the basis for dialogue that would use a global bioethics framework; some of this abundant literature has been briefly referenced in this chapter. With respect to the *national context*, the realities of systems in place (e.g., education, health, transportation) would be considered and (as the EDEN document suggests), both local and external planners and emergency response volunteers would need to be cognizant of the systems. Here, too, divergent values within the country or countries addressing CBRNE would need to be considered, of course. Simply acknowledging those differences within the context of a discourse of ethical concerns, while also seeking commonality, can facilitate dialogue. *International human rights laws and practices* would need to be considered, including with respect to trade-offs that might be necessary in terms of local norms and practices (e.g., regarding age groups).

In addition, as is suggested in the European Commission's PBIED Ethics Guideline Document, basic principles of bioethics developed by the UN should be considered. In 2009, UNESCO promulgated a Universal Declaration on Bioethics and Human Rights that, while not specifically devised to address ethical issues related to CBRNE, includes principles that apply to these circumstances and that were ostensibly based on consideration of divergent religious, cultural, and national beliefs and values (UNESCO 2009). Applying the global health ethics framework as a point of departure for discourse regarding its practical application would be useful. Let us consider the UNESCO principles and their application to an ethical concern regarding CBRNE; that is: significant technological changes in and increasing use of "big data" to prevent and address the consequences of CBRNE threats. Three of the UNESCO principles that apply are:

> Article 9—Privacy and confidentiality
> The privacy of the persons concerned and the confidentiality of their personal information should be respected. To the greatest extent possible, such information should not be used or disclosed for purposes other than those for which it was collected or consented to, consistent with international law, in particular international human rights law.
>
> Article 10—Equality, justice and equity
> The fundamental equality of all human beings in dignity and rights is to be respected so that they are treated justly and equitably.
>
> Article 11—Non-discrimination and non-stigmatization
> No individual or group should be discriminated against or stigmatized on any grounds, in violation of human dignity, human rights and fundamental freedoms.

The principle of privacy and confidentiality has become increasingly complex in light of the use of big data. Indeed, this is a quintessential example of one of the domains of the framework—*international (and national) law and policies*—that is being outdistanced by technological advances. The urgency of the need for data and information to prevent a threat would need to be considered in light of the need to protect privacy, but this in turn must be based on international and national and local human rights laws and practices, national systems in place to protect privacy, and political and social systems and processes in place to ensure a balance between protection of individual rights and the responsibility to protect society. Responding to a CBRNE event requires determining the need for data to inform immediate and longer-term responses that also must take into account two other framework domains—national laws and information systems in place and the national context. This would include the likelihood of accuracy and relevance for decision-making and trust in both national and local leaders and in outsiders and its corollary, fear of reprisal for information provided; and existence or lack of systems for feedback regarding decisions made based on the information (e.g., equitable allocation of resources).

The way in which CBRNE planners and responders could use the Global Bioethics Framework for CBRNE is contingent on their role and responsibility. Policymakers and planners should could use the framework as a basis for long-term planning at the international, regional and country level (including considering values and ethical principles of diverse countries) and for engagement with a wide range of stakeholders, including community members. They could also use it to rapidly plan for response to a CBRNE emergency. However, to do so requires the involvement of multiple organizations and communities in the processes to explore a universal set of ethical principles for CBRNE, recognizing similarities and building on these but taking into account diversity. This requires public and private sector collaboration, and participation on the part of diverse communities at all levels of society.

Emergency responders should be trained in bioethics related to CBRNE generally, using the framework as one of the tools in the training. While it is impractical to assume that there would be adequate time for emergency responders who are not from a country to which they are sent for a CBRNE emergency, such advance training in bioethics related to CBRNE generally would help to ensure that they are prepared to address the urgent value-laden decisions they may face.

5.4 Conclusion

The development and application of a Global Bioethics Framework to CBRNE in practice would, in the view of this author, enhance the dialogue regarding the complex, challenging issues inherent in the field. What are the arguments for and against doing using such a framework? Arguments in favour of such a broad framework include an increasing awareness of comparable core moral values (as distinct from cultural practices) with regard to, for example, good will; the need to have a common set of guidelines to enable an effective and ethical response to disasters; and the burgeoning dialogue across faiths and cultures. The arguments against such a global framework include the concern that in practice it might impede the ability of responders to be situational in a positive sense; that is, to consider national circumstances and culture; and the possibility that there would be lack of clarity in the interest of generalizability. In practice with respect to responses to CBRNE, this could include confusion regarding sanctity and (relative) value of human life, and equity regarding resource distribution.

For the foreseeable future, there will be actual or potential CBRNE events that require, over the short and long term, that societies respond to urgent human needs. We must take responsibility to ensure that our common values are considered in the development of national and international policies and public instruments and programs to address them at the same time that local culture and contexts are fully considered.

References

Andorno, Roberto. 2007. Global bioethics at UNESCO: In defence of the universal declaration on bioethics and human rights. *Journal of Medical Ethics* 33: 150–154.

Beauchamp, Thomas, and James Childress. 1979. *Principles of bioethics*. New York: Oxford University Press.

Benatar, Solomon R., Abdallah S. Daar, and Peter A. Singer. 2003. Global health ethics: The rationale for mutual caring. *International Affairs* 79 (1): 107–138.

Caton, Jeffrey. 2015. *Autonomous weapon systems: A brief survey of developmental, operational, legal, and ethical issues*. Carlisle: Strategic Studies Institute and U.S. Army War College Press.

Council for International Organizations of Medical Sciences. 1993. *International ethical guidelines for biomedical research involving human subjects*. Geneva: CIOMS.

———. 2016a. *Evidence synthesis and meta-analysis for drug safety*. Geneva: CIOMS.

———. 2016b. *International ethical guidelines for health-related research involving humans*. Geneva: CIOMS.

———. 2018. *CIOMS guide to vaccine safety communication*. Geneva: CIOMS.

Drane, James. 2002. What is bioethics? A history. In *Interfaces between bioethics and empirical social sciences*, ed. Fernando Lolas and Lorenzo Agar, 15–31. Buenos Aires: Regional Program on Bioethics of the Division of Health and Human Development of the PAHO/WHO.

EDEN Consortium. 2014. End-user driven demo for CBRNe report on ethical issues of prevention, preparedness and detection phase. *EDEN Consortium* 81–2.

El Setouhy, Maged, Tsiri Agbenyega, Francis Anto, Christine A. Clerk, Kwadwe A. Koram, Michael English, et al. 2004. Moral standards for research in developing countries: From "reasonable availability" to "fair benefits.". *The Hastings Center* 34 (3): 17–27.

Farmer, Paul, and Nicole Campos. 2004. Rethinking medical ethics: A view from below. *Developing World Bioethics* 4 (1): 17–41.

Feinholz, Dafna. 2015. *UNESCO and bioethics: Make bioethics everyone's business*. UNESCO. www.unesco.org/shs/bioethics. Accessed 19 June 2017.

Goldsand, Gary, Zahava R.S. Rosenberg, and Michael Gordon. 2001. Bioethics for clinicians: 22. Jewish bioethics. *CMAJ* 164 (2): 219–222.

Jillson, Irene A. 2011. Global health ethics: Toward common principles and their practical application to health research, clinical care, and systems management. Keynote address: *Second International Congress on Bioethics*. Tehran, Iran.

Karenga, Maulana. 2003. *MAAT: The moral ideal in ancient Egypt*. New York: Routledge.

Kirby, Michael. 2015. Article 2: Aims. In The UNESCO universal declaration on bioethics and human rights: Background, principles and application, ed. Henk A.M.J. ten Have and Michele S. Jean. Paris: UNESCO, 67–80.

London, Alex J. 2005. Justice and the human development approach to international research. *The Hastings Center* 35: 24–37.

McCarthy, Joan. 2003, December 8. Principlism or narrative ethics: Must we choose between them? *Medical Humanities*. 65–71. http://mh.bmj.com/Med Humanities. Accessed 20 Oct 2018.

O'Neill, Onora. 2002. Public health or clinical ethics: Thinking beyond borders. *Carnegie Council for Ethics & International Affairs* 16 (2): 35–45.

Pellegrino, Edward D. 2005. Decision at the end of life: The use and abuse of the concept of futility. *Practical Bioethics* 1 (3): 3–6.

Rennie, Stuart, and Bavon Mupenda. 2008. Living apart together: Reflections on bioethics, global inequality and social justice. *Philosophy, Ethics, and Humanities in Medicine* 3: 25.

Sakamoto, Hyakudai. 1999. Towards a new 'global bioethics. *Bioethics* 13 (3–4): 191–197.

Shomali, Mohamamd Ali. 2008. Islamic bioethics: A general scheme. *Journal of Medical Ethics and History of Medicine* 1: 1.

Singer, Peter A., and Soloman R. Benatar. 2001. Beyond Helsinki: A vision for global health ethics. *BMJ* 322: 748–788.

Spiegel, Allen D. 1997. Hammurabi's managed health care — circa 1700 B.C. *Managed Care*. https://www.managedcaremag.com/archives/1997/5/hammurabis-managed-health-care-circa-1700-bc. Accessed 20 Dec 2017.

Stanciugelu, Irina. 2015. *Deliverable D5.1 PBIED ethics guideline document*. European Commission. https://www.subcop.eu/SUBCOP%20D5.1_PBIED%20Ethical%20Guideline_v7-Final%20(4).pdf. Accessed 15 June 2018.

Stevens, M.L. Tina. 2014. The history of bioethics: Its rise and significance. *Reference Module in Biomedical Sciences*. https://doi.org/10.1016/B978-0-12-801238-3.00175-6.

Tai, Michael Cheng-Tek. 2008. The way of Asian bioethics. *Asian Bioethics Review* 1: 15–23.

Tangwa, Godfrey B. 1999. Globalisation or westernization? Ethical concern in the whole bio-business. *Bioethics* 13 (3–4): 218–226.

UNESCO. 2009. In *The UNESCO universal declaration on bioethics and human rights: Background, principles and application*, ed. Henk A.M.J. ten Have and Michele S. Jean. Paris: UNESCO.

Chapter 6
Justice and Good Governance in Nuclear Disasters

Behnam Taebi

Abstract This chapter has two goals. First, I will present *justice* as an overarching notion for addressing the ethical issues of disaster governance. In different stages of disaster mitigation, preparation, response and recovery, there are three justice issues at play, namely distributive justice, recognition of those affected and procedural justice. Then second goal of this chapter is to spell out implications of procedural justice for disaster governance. More specifically, I argue that *good governance* of disasters requires that we at least warrant (i) *the quality of information* (i.e. transparency, credibility of knowledge and information and trust) and (ii) the fairness in stakeholders engagement (i.e. inclusiveness and due process in participation). Could *citizen science* contribute to more reliable and *transparent* information and more *accountable* local and national governments and corporations? Would regional (i.e. supranational) disaster governance better ensure the credibility and transparency of information?

The chapter will focus on disasters that involve radiological risk such as the Fukushima-Daiichi disaster in 2011, but the rationale of the argument is more broadly applicable. Similar to discussions on good governance of risk, I argue that the ambition of *good disaster governance* should be to provide a conceptual and normative framework to deal with the complexity of knowledge and information and to ensure a fair process during all stages of disaster governance.

Keywords Disaster governance · Risk governance · Distributive justice · Procedural justice · Ethics of radiation · Nuclear ethics

Society often stands to benefit greatly from all the new technologies that are introduced but new and significant risks can also emerge. These risks bring a wealth of

---

B. Taebi (✉)
Delft University of Technology, Delft, The Netherlands

Harvard University, Cambridge, MA, USA
e-mail: B.Taebi@tudelft.nl

D. P. O'Mathúna, I. de Miguel Beriain (eds.), *Ethics and Law for Chemical, Biological, Radiological, Nuclear & Explosive Crises*, The International Library of Ethics, Law and Technology 20, https://doi.org/10.1007/978-3-030-11977-5_6

ethical issues which, in their turn, add to the complexity of governance. A key ambition of risk governance has therefore been to acknowledge and to identify these ethical issues (e.g. Asveld and Roeser 2009; Hansson 2015; Taebi 2017) and to provide a normative basis for responsibly dealing with such uncertain and complex risks (van Asselt and Renn 2011). One type of risk that is particularly difficult to govern is nuclear risk, or even the radiological risks associated with nuclear energy technologies. The problems that culminated in the Fukushima-Daiichi disaster were not only traceable to an insufficiency of technical measures, but also to a lack of profound understanding of socio-ethical issues, such as a lack of transparency and a proper safety culture (Ahn et al. 2015; Figueroa 2016).

In this chapter I shall focus on the governance challenges of nuclear risks in a disaster setting. By drawing parallels with discussions on risk governance I shall argue that possessing *good* disaster governance in conjunction with nuclear energy technologies would require us to understand and include the societal and ethical issues associated with nuclear risk. These normative issues will be spelled out for nuclear technology by focusing on distributive and procedural justice.

The chapter is organized as follows. In Sect. 6.1, the focus will be on the particular (notably morally laden) features of nuclear risks which warrant special focus, especially in the light of disaster governance. Section 6.2 elaborates on the notion of good nuclear disaster governance. Section 6.3 will review several International Commission on Radiological Protection (ICRP)[1] publications which serve to indicate that considerations concerning justice and radiological protection are at least four decades old. That Section will also elaborate on the specific considerations of distributive and procedural justice in relation to nuclear risk disaster governance. In Sect. 6.4, I will present my conclusions.

6.1 What Justifies a Specific Focus on Radiation Risks?

The first question is whether we should specifically focus on radiological risks emanating from nuclear energy technologies. I believe that such specific focus is justified because radiation risks are different to other types of technological risks in several respects. Firstly, as with other types of technological risks, radiation risks bring great uncertainty, but these uncertainties are very difficult to manage because there is no such thing as a safe level of radiation exposure. Secondly, but closely related to the same issue of uncertainty, radiation exposure is not a one-off occurrence; it is the accumulation of radiation that can have an impact on human health;

[1] There is also a recent discussion, started by the International Commission on Radiological Protection (ICRP), on the ethical foundation of the system of radiological protection. The findings will culminate in the publication of a new guideline publication that will spell out these ethical issues and help practitioners by providing the relevant tools and procedures that could be used (ICRP 2018). While the publication is about radiological risks in general, I believe that there are some interesting lessons to be learned for disaster governance in the case of nuclear risk.

this impact might only manifest itself in a human after a very long time. Thirdly, (and somewhat on the same topic) radiation could remain in the environment for a very long time thus presents a long-term legacy for many generations to come. One need only think of the nuclear disasters areas around Chernobyl or Fukushima-Daiichi that have become virtually permanently uninhabitable. Generally therefore, any major nuclear disaster can be said to have devastating impacts far beyond generational borders which do, indeed, extend into the very long-term future. Fourthly, nuclear risk can also extend beyond national borders, either as a risk that is produced in one country and could reach neighboring countries or as a type of risk that will be produced together with a number of countries. This requires further elucidation.[2]

Nuclear risk could be generated by several countries if one country decides to dispose of nuclear waste originating from another country. While the current focus in nuclear waste disposal policy is on national disposal areas, there are several developments for joint disposal that are worth mentioning. The first concerns Europe where different national nuclear waste organizations are exploring the possibility of a European nuclear waste repository (Salzer et al. 2012) and the second pertains to a recent decision by the Government of South Australia to consider the possibility of hosting an international disposal location for nuclear waste in that region (Scarce 2016). Indeed, such internationally produced nuclear risk would give rise to complex justice-related implications, both in distributive and procedural respects (e.g. Taebi 2012a, 2018).

6.2 Why Emphasize Good Governance?

While the notion of good governance has been most commonly discussed in relation to the need for accountability, transparency and rule of law in developing countries, Rothstein (2011) argues that various cases of *bad governance*, such as the global financial crisis, showed the relevance of this notion for all developed counties. The nuclear disaster in Fukushima-Daiichi is clearly an example of bad or poor governance on the part of policy-makers (including the Japan Nuclear Regulatory Authority and the Tokyo Electric Power Company (TEPCO), both regarding how the accident came about and how it was responded to (Lochbaum et al. 2014). What is often neglected in discussions on good governance is the normative dimension, or the question as to what constitutes 'good' (Rothstein 2012; Taebi 2017). The Sendai Framework for Disaster Risk Reduction also emphasizes the need to "strengthen good governance in disaster risk reduction strategies at the national, regional and global levels" (UN 2015, 10). However, what precisely good governance entails remains unclear.

[2]The same reasoning has been presented to justify a specific focus on the 'ethics of nuclear energy; see for instance (Taebi et al. 2012; Taebi and Roeser 2015).

I argue that discussions on the good governance of disasters should aim at providing a normative basis for how to deal with complex, uncertain, intergenerational and international risks. These normative issues are undoubtedly the ones that those involved in disaster management are already dealing with on a daily basis. What I would like to emphasize is that we should explicitly acknowledge the normativity of such questions like, for instance, the question as to which groups of people should be helped when resources are limited and choices need to be made. This example will be elaborated on in the discussion on issues of distributive justice.

6.3 Justice and Radiation Protection

The principle of *equity*[3] in good governance emphasizes the fact that the interests of all members of the society need to be respected and taken into account. Considerations of justice have played a part in radiation related issues for about four decades. In a seminal publication by the ICRP in 1977, the three principles of radiological protection – Justification, Optimization and the Dose Limit Principle – that appeal to a justice rationale were presented. The Justification Principle, for instance, dictates that any activity that alters the radiation exposure situation should do more good than harm (ICRP 1977); this is a consequentialist interpretation of how to proceed when radiation exposure is justified. Justification is indeed something other than justice, but the principle aims at a *fair* balance of the burdens and benefits. Likewise, the Optimization Principle in radiological protection recommends that when people are exposed to radiation, the level should be kept as low as is reasonably possible, taking all societal and economic factors into account. This, again, is a consequentialist aggregative argument about the exposure to which people can be subjected. Even when there is *justifiable* benefit, radiation practitioners need to *optimize*, that is, minimize radiation as far as circumstances allow. The third principle of radiological protection is the Dose Limit Principle that states that individual exposure should not exceed the levels recommended by the ICRP. The Dose Limit is an implicit acknowledgement of potential inequities that could arise from the Optimization principle. "When the benefits and detriments do not have the same distribution through the population, there is bound to be some inequity. Serious inequity can be avoided by the attention paid to the protection of individuals" (ICRP 1991, Par. 101). It would go beyond the scope of this short piece to engage with the fundamental ethical issues that these principles raise[4]; the reason why I am including this quotation is to show that justice in relation to radiation has been a relevant subject for several decades. Also in the most recent publication of the ICRP, justice is mentioned as one of the four key fundamental values that underlie the system of radiological protection. More specifically, the ICRP argues that radiological

[3] For the purpose of this chapter I will not distinguish between principles of equity and justice. In this chapter they will be taken to mean the same.

[4] See for instance: (Gardiner 2008; Oughton and Hansson 2013; Valentin 2013).

protection should firstly aim "to reduce *inequities* in the distribution of individual exposures in situations where some individuals could be subject to much more exposure than others" and should secondly "ensure that exposures do not exceed the values beyond which the associated risk is considered as not *tolerable* given a particular context" (ICRP 2018, 30–31 emphasis added by me). In addition to distributive justice, this publication mentions (but does not discuss) restorative justice (or: compensation for losses) and procedural justice (or: due process in decision-making). In the following subsections I will now go on to briefly review the distributive and procedural justice issues that are associated with the different stages of the disaster cycle in a nuclear incident.[5]

6.3.1 *Distributive Justice*

Any activity involving radiation is primarily proposed to bring some benefit to some,[6] but it could also pose radiological risk to others, all of which gives rise to questions on how to distribute risks and benefits. While the benefits often tend to extend to a very large group of people, the risks are usually much more localized. This is an important ethical issue associated with exposure to radiation risk. The central goal of distributive justice is to ensure that risks and benefits are distributed in a *just* manner. Here there are at least three questions that give rise to fundamental ethical quandaries, namely what is the *shape*, *currency*, and *scope* of distribution. In other words, which patterns of distribution do we prefer (shape), what is it that we wish to distribute (currency) and to whom does this distribution relate (scope) (building on Page 2006). Regarding the shape, different distributional principles could be followed. A utilitarian principle, for instance, would prescribe a distribution that could maximize the utility of all, while a Rawlsian model of distribution would aim to help the least well off. What complicates this issue is the fact that radiation workers, people in the vicinity of the reactor (who could be affected by the accumulation of small doses of radiation) and people outside the direct impact zone (who will only be affected in the event of a major accident) are exposed to different radiation levels. To what extent is this morally justified? The *currency* question, or the question as to what it is that we wish to *distribute*, could be answered in different ways too; vulnerabilities, resources, capabilities or resilience are the possible units or currencies of distribution that have been proposed (Doorn 2015). Both the shape and the currency questions become more complex when we consider the question of *scope* or, to whom this distribution relates, particularly when we need to consider

---

[5] The distinction mentioned here, but also my account of why justice matters in disaster governance is somewhat building on the ongoing discussions on Energy Justice, which in turn stem from the literature on environmental justice. The Energy Justice scholarship propose a tripartite model of distribution, procedure and recognition (e.g. Jenkins et al. 2016).

[6] This is in line with the ICRP's *Justification Principle* which requires that any decision that alters the radiation exposure situation should do more good than harm (ICRP 2007).

temporal distribution and the associated intergenerational justice questions. What is it that we can pass on to future generations and what is the moral justification for that (e.g. Taebi 2012b; Kermisch and Taebi 2017)?

It is not my intention to spell out and reflect on possible answers to both the shape and the currency questions. Indeed, in actual disaster management, these questions are always dealt with, but often implicitly. For instance, in disaster management there are several dilemmatic questions such as the following: when resources are limited, should we first help 10 people in dire need and in remote areas or 100 people who also need (less urgent) care and in closer proximity? Indeed, a utilitarian and a Rawlsian would respond to this question (drastically) differently. In good disaster governance such questions do, however, deserve to be discussed explicitly and together with the different stakeholders involved. That brings me to the next issue of procedural justice.

6.3.2 Procedural Justice

The procedural justice issue evidently coincides with the central principles of *good governance* that aims to encourage a participatory consensus-based decision-making stance. It is information that plays a key role here. The transparency principle, for instance, seeks to guarantee open access to information for all stakeholders. What matters most is how to ensure *the quality of information and information transfer*, that is, the access of all stakeholders to understandable and reliable information. Before moving on to discussions on how this quality of information could be improved, let me first draw an analogy between risk and disaster governance that will help me to illustrate the central problematic issue in hand. It has been argued that technocratic approaches to risk governance – based on the notion that scientists and technical experts provide the knowledge and expertise upon which the policy is formulated and the public must act – are not helpful, because they tend to ignore important social and ethical aspects of risk, such as uncertainties (van Asselt and Renn 2011; Kaliarnta et al. 2014; Taebi et al. 2016). Similarly, I argue that technocratic approaches to disaster governance are errant because they suffer from the following two deficiencies. In the first place they neglect the issues of controversies regarding scientific knowledge as well as disagreements among experts. The overabundance of evacuation maps after the Fukushima-Daiichi disaster with divergent and sometimes contradictory recommendations is a good example of such disagreement. In the second place they ignore the complexity of scientific knowledge, including all the relevant uncertainties. Acknowledging these uncertainties, the International Commission on Radiological Protection recommends upholding the value of *prudence* in radiological protection practices where decision-making is needed without full knowledge of the consequences.[7] In the remainder of this

[7] Prudence is one of the values that the ICRP is promoting as a key value in the system of radiological protection. This is part of the ongoing discussion on stipulating key values, but prudence has been a value discussed in previous ICRP-publications such as in ICRP (2007).

sub-section two ways of ensuring the quality of information will be put forward first, by providing additional (sometimes unconventional) sources of information and second, by deploying multinational overseeing mechanisms. These ideas will be reviewed while simultaneously proposing future research possibilities designed to investigate their viability.

One way to improve the quality of information is offered by *citizen science*. It is based on the notion of encouraging broad groups of citizens to become more involved in decision-making (which is one of the primary goals of the good governance) while also aiming to enable citizens to contribute to "collecting, categorizing, transcribing, or analyzing scientific data" (Bonney et al. 2014, 1436). There has already been some evidence that "volunteers perform almost as well as professionals in some areas" but there are some concerns about data quality (Crall et al. 2011, 433). Good experience has already been gained with volunteers who used an Open Source Network to collect radiation data around the crippled Fukushima-Daiichi reactors.[8] Indeed, implementing citizen science as a source of information would pose several ethical and epistemological challenges but if successful this *unconventional* method could provide a valuable secondary source of information. The empirical question that would need further investigation would be: to what extent could citizen science contribute to more *accountability* on the part of local or national governments and corporations by delivering more open-source data on controversial issues?

Another important issue linked to procedural justice is the fact that some stakeholders (who could be affected by the consequences of a nuclear accident) might be in the neighboring countries. This calls for supranational nuclear risk governance in an international or perhaps multinational or regional setting. Also the Sendai Framework calls for more supranational and regional collaboration. It is expected that such a proposal would run into difficulties when it comes to the matter of national sovereignty: a state would have to give other states access to important information and accept an overseeing mechanism that goes beyond the state itself. The issue of global, regional or supranational governance of nuclear safety has already been the subject of various inquiries since the Fukushima-Daiichi disaster. Historically, after a major disaster, there is more inclination among nation states to move towards more nuclear safety globalization (e.g. Bunn and Heinonen 2011), but the fundamental questions is whether nuclear safety should be increased "by accident or by design" (Taebi and Mayer 2017)?[9] The empirical question for future research could therefore be: to what extent could regional disaster governance do a better job of ensuring the credibility and transparency of information? What ingredients (through monitoring, verification and enforcement) should such supranational governance include?

---

[8] This is based on an Open Source Network known as Safecast, which enabled citizens with no knowledge about radiation to contribute to the collection of data by using do-it-yourself kits. http://voices.nationalgeographic.com/2016/02/13/how-citizen-science-changed-the-way-fukushima-radiation-is-reported

[9] This has important implications for the destining of nuclear policy and nuclear technologies such as reactors (Taebi and Kloosterman 2015).

6.4 Conclusions

In this contribution I have argued that the discussions on the good disaster governance of nuclear risk should aim at providing a normative basis to deal with complex, uncertain, intergenerational and international risks. This framework should at least address fundamental questions of justice. Distributive justice involves specifying the shape, currency and scope of justice for different groups of people while procedural justice is about ensuring the quality of information and increasing accountability for all state and non-state actors. As regards distributive justice, questions will be addressed such as: why is it morally justifiable for radiation workers, people in the vicinity of reactors and others to be exposed to different levels of radiation? What is the proper unit of *distribution*? How should we deal with international and intergenerational distribution? As regards procedural justice, the role of information is becoming increasingly relevant. How can we increase the quality of information and information transfer? Could *citizen science* contribute to more reliable and *transparent* information and more *accountable* local and national governments and corporations? Would regional (i.e. supranational) disaster governance better ensure the credibility and transparency of information?

These are indeed conceptual questions. Future research should focus on the empirical investigation behind these questions, but that is an endeavor that is certainly worth undertaking, since lack of profound understanding of societal and ethical issues could potentially lead to disasters or to improper responses after a disaster.

References

Ahn, J., C. Carsen, M. Jensen, K. Juraku, S. Nagasaki, and S. Tanaka, eds. 2015. *Reflections on the Fukushima Daiichi nuclear accident – Toward social-scientific literacy and engineering resilience*. Heidelberg: Springer Open.

Asveld, L., and S. Roeser, eds. 2009. *The ethics of technological risk*. London: Earthscan.

Bonney, R., J.L. Shirk, T.B. Phillips, A. Wiggins, H.L. Ballard, A.J. Miller-Rushing, and J.K. Parrish. 2014. Next steps for citizen science. *Science* 343: 1436–1437.

Bunn, M., and O. Heinonen. 2011. Preventing the next Fukushima. *Science* 333: 1580–1581. https://doi.org/10.1126/science.1209668.

Crall, A.W., G.J. Newman, T.J. Stohlgren, K.A. Holfelder, J. Graham, and D.M. Waller. 2011. Assessing citizen science data quality: An invasive species case study. *Conservation Letters* 4: 433–442. https://doi.org/10.1111/j.1755-263X.2011.00196.x.

Doorn, Neelke. 2015. Resilience indicators: Opportunities for including distributive justice concerns in disaster management. *Journal of Risk Research*: 1–21.

Figueroa, P. 2016. Nuclear risk governance in Japan and the Fukushima triple disaster: Lessons unlearned. In *Disaster governance in an urbanising Asia*, ed. M.A. Miller and M. Douglass, 263–282. Singapore: Springer.

Gardiner, S. 2008. Why we need more than justification in the ethics of radiological protection: A view from outside. In *Ethics and radiological protection*, ed. G. Eggermont and B. Feltz, 97–111. Louvain-la-Neuve: Academia-Bruylant.

Hansson, S.O. 2015. Nuclear energy and the ethics of radiation protection. In *The ethics of nuclear energy: Risk, justice and democracy in the post-Fukushima era*, ed. B. Taebi and S. Roeser, 17–34. Cambridge: Cambridge University Press.

ICRP. 1977. *Recommendations of the international commission on radiological protection. Publication 26*. Annals of the ICRP 1 (3). Oxford: Pergamon Press.

———. 1991. *1990 Recommendations of the international commission on radiological protection. ICRP Publication 60*. Annals of the ICRP 21 (1–3). Oxford: Pergamon Press.

———. 2007. *The 2007 recommendations of the international commission on radiological protection. Publication 103*. Annals of the ICRP 37 (2–4). Oxford: Elsevier.

———. 2018. *Ethical foundations of the system of radiological protection. Publication 138*. Annals of the ICRP 47 (1). Sage.

Jenkins, K., D. McCauley, R. Heffron, H. Stephan, and R. Rehner. 2016. Energy justice: A conceptual review. *Energy Research & Social Science* 11: 174–182.

Kaliarnta, S., M. Hage, and S. Roeser. 2014. Involving stakeholders in risk governance. In *Trade, health and the environment: The European Union put to the test*, ed. M. Everson, E. Vos, and M.B.A. van Asselt, 235–253. London: Routledge.

Kermisch, C., and B. Taebi. 2017. Sustainability, ethics and nuclear energy: Escaping the dichotomy. *Sustainability* 9: 446.

Lochbaum, D., E. Lyman, S.Q. Stranahan, et al. 2014. *Fukushima: The story of a nuclear disaster*. New York/London: The New Press.

Oughton, D., and S.O. Hansson. 2013. *Social and ethical aspects of radiation risk management*. Amsterdam: Elsevier.

Page, E. 2006. *Climate change, justice and future generations*. Cheltenham: Edward Elgar Publishing.

Rothstein, Bo. 2011. *The quality of government: Corruption, social trust, and inequality in international perspective*. Chicago: University of Chicago Press.

———. 2012. Good governance. In *The Oxford handbook of governance*, ed. David Levi-Faur, 143–154. Oxford: Oxford University Press.

Salzer, P., J. Pritrsky, A. Mrskova, and P. Richardson. 2012. *The status of multinational waste management solutions*. Deliverable 3.3. of the IPPA (Implementing Public Participation Approaches in Radioactive Waste Disposal).

Scarce, K. 2016. *Nuclear fuel cycle Royal Commission report*. Adelaide: Government of South Australia.

Taebi, B. 2012a. Multinational nuclear waste repositories and their complex issues of justice. *Ethics, Policy & Environment* 15: 57–62.

———. 2012b. Intergenerational risks of nuclear energy. In *Handbook of risk theory. Epistemology, decision theory, ethics and social implications of risk*, ed. S. Roeser, R. Hillerbrand, P. Sandin, and M. Peterson, 295–318. Dordrecht: Springer.

———. 2017. Bridging the gap between social acceptance and ethical acceptability. *Risk Analysis* 37: 1817–1827.

———. 2018. How to bridge the gap between social acceptance and ethical acceptability: A Rawlsian approach. In *Environmental health risks. Ethical aspects*, ed. F. Zölzer and G. Meskens, 45–64. Abingdon: Earthscan from Routledge.

Taebi, B., and J.L. Kloosterman. 2015. Design for values in nuclear technology. In *Handbook of ethics, values, and technological design: Sources, theory, values and application domains*, ed. J. Van den Hoven, P. Vermaas, and I. Van de Poel, 805–829. Dordrecht: Springer.

Taebi, B., and M. Mayer. 2017. By accident or by design? Pushing global governance of nuclear safety. *Progress in Nuclear Energy* 99: 19–25.

Taebi, B., and S. Roeser, eds. 2015. *The ethics of nuclear energy: Risk, justice and democracy in the post-Fukushima era*. Cambridge: Cambridge University Press.

Taebi, B., S. Roeser, and I. Van de Poel. 2012. The ethics of nuclear power: Social experiments, intergenerational justice, and emotions. *Energy Policy* 51: 202–206.

Taebi, B., A. Correljé, E. Cuppen, E. van de Grift, and U. Pesch. 2016. Ethics and impact assessments of large energy projects. In *Proceedings of the IEEE.Ethics 2016*. Vancouver.
UN. 2015. *Sendai framework for disaster risk reduction 2015–2030*. Sendai: United Nations (UN).
Valentin, J. 2013. Radiation risks and the ICRP. In *Social and ethical aspects of radiation risk management*, ed. D. Oughton and S.O. Hansson, 17–32. Amsterdam: Elsevier.
van Asselt, M.B.A., and O. Renn. 2011. Risk governance. *Journal of Risk Research* 14: 431–449.

Chapter 7
Post-Lisbon Developments in EU Crisis Management: The Integrated Political Crisis Response (IPCR) Arrangements

Agnieszka Nimark

Abstract This chapter describes the new developments taking place in European Union (EU) crisis management, in particular post-Lisbon Treaty arrangements related to the EU response to major crises. The implementation of the Solidarity Clause and adoption of the Integrated Political Crisis Response (IPCR) arrangements indicates the Union's determination to systematize its decision-making procedures in response to crisis situations. This chapter explains first the main impediments for the use of the IPCR predecessor – Emergency and Crisis Coordination Arrangements (CCA) and the long consultation process leading to the adoption of the IPCR. Subsequently, the chapter explains how the new institutional set-up for crisis response at the high political level works and how the EU Member States and institutions interact in the event of major crises. The analysis focuses on the legal framework of decision-making at the EU level and on the operational aspects aimed at strengthening the EU's capacity to collect and analyse real time information necessary for strategic decision-making. First activation of the IPCR is discussed in the context of the EU response to the refugee and migration crisis.

Keywords Crisis management · Crisis response · CCM · Integrated Political Crisis Response (IPCR) · Solidarity clause

7.1 Introduction

The entry into force of the Lisbon Treaty brought some intriguing developments regarding the EU's role in crisis management. On the one hand, the Treaty explicitly reaffirmed that '*national security remains the sole responsibility of each Member*

A. Nimark (✉)
CIDOB - Barcelona Centre for International Affairs, Barcelona, Spain

Judith Reppy Institute for Peace and Conflict Studies, Cornell University, Ithaca, NY, USA
e-mail: an355@cornell.edu

D. P. O'Mathúna, I. de Miguel Beriain (eds.), *Ethics and Law for Chemical, Biological, Radiological, Nuclear & Explosive Crises*, The International Library of Ethics, Law and Technology 20, https://doi.org/10.1007/978-3-030-11977-5_7

State' (Article 4 of the Treaty on European Union).[1] On the other hand, while reiterating Member States' primary responsibility for the management of crises within their territory, the Treaty also introduced some new instruments, notably the so-called *solidarity*[2]*clause*, for improving the coherence and complementarity of Union and Member States' actions in crisis management. Pursuant to Article 222(1) of the Treaty on the Functioning of the European Union (TFEU), '*the Union and the Member States are to act jointly in a spirit of solidarity if a Member State is the object of a terrorist attack or the victim of a natural or man-made disaster*'. As a result, one might have the impression of an unclear division of crisis management competences between national and European levels. However, the adoption of the Council Decision *on the arrangements for the implementation by the Union of the solidarity clause* in June 2014 helped to clarify the competences of the EU and the Member States in the event of a major crisis.[3]

Even though the Council Decision does not concern the implementation of the solidarity clause by the Member States,[4] it does regulate how the Member States are to coordinate between themselves within the Council in a crisis situation. The central role of the Council in the event of the invocation of the solidarity clause, as well as in other types of emergencies, is also underpinned by the approval in June 2013 of the *EU Integrated Political Crisis Response* (IPCR) arrangements.[5] The IPCR involves a variety of the EU bodies and tools as support of the decision-making process. In addition, the new arrangements can be activated in crisis situations independently of the invocation of the solidarity clause. In light of these new developments, the objective of this paper is two-fold: first, to provide an analysis of the Lisbon Treaty's legal and institutional framework with the aim to explain who is in charge at the EU level when a major crisis affects one or more European states; second, to examine the legal and operational aspects of the new IPCR arrangements.

---

[1] Similarly Article 72 of the Treaty on the Functioning of the European Union states in regard to Title V – Area of Freedom, Security and Justice – that: *This Title shall not affect the exercise of the responsibilities incumbent upon Member States with regard to the maintenance of law and order and the safeguarding of internal security.*

[2] On *solidarity* as a European concept, See: Raspotnik, Andreas, Marine Jacob, and Laura Ventura. 2012. The Issue of solidarity in the European Union, *TEPSA Brief*. August 8. http://www.tepsa.eu/download/TEPSA Policy Paper The issue of solidarity in the European Union.pdf Accessed 3 October 2018.

[3] Council Decision 2014/415/EU of 24 June 2014 on the arrangements for the implementation by the Union of the solidarity clause, OJ L192/53, 1 July 2014.

[4] According to Article 222(2) and Declaration (No 37) of the Treaty on the Functioning of the European Union: *a Member State can choose the most appropriate means to comply with its own obligation towards another Member State*.

[5] On the 25 of June 2013 the Council of General Affairs approved the document 10708/13 regarding Finalisation of the CCA review process: the EU Integrated Political Crisis Response (IPCR) arrangements. http://www.consilium.europa.eu/register/en/content/out/?&typ=ENTRY&i=ADV&DOC_ID=ST-10708-2013-INIT. Accessed 3 October 2013.

7.2 The EU's Role in Crisis Management

Originally, the European Union (EU), or rather the European Communities (EC), were not intended to deal with security and safety matters, such as defence or civil protection, as these responsibilities were part of the sovereign competences of the individual Member States. However, as early as the 1970s, in the aftermath of serious environmental accidents such as the dioxin cloud caused by the *Seveso* chemical plant in Italy (1976), or the breaking apart of the oil tanker *Amoco Cadiz* near the French coast (1978), the Member States agreed for the first time that joint action was necessary to prevent similar and potentially more disastrous accidents in the future.[6] As a result of these events, a more general idea was also born that the EC might play a role in helping Member States requiring expert assistance and critical resources in the management of a crisis (Boin et al. 2013).[7]

Later on, the prospects of disasters that could overwhelm a single member state, e.g. the series of forest fires in Italy in the 1980s, gave rise in 1987 to the first decision to allow the Commission to begin a network for national civil protection officers.[8] Eventually, the EU's Civil Protection Mechanism was established based on an idea that large-scale natural disasters that overwhelm a member state require collaboration among assisting countries.[9] In a similar way, other large-scale accidents and crises had a strong impact on EU policy developments related to crisis and disaster management. For example, the spreading over Europe of a radioactive cloud caused by Chernobyl (1986) influenced adoption of new regulations in areas such as health, trade and emergency response[10]; the EU's difficulties in dealing with Bovine spongiform encephalopathy (BSE, or 'mad cow disease') and its connection with Creutzfeld-Jakob disease led eventually to establishment of the Directorate General (DG) for Health and Consumer Protection including crisis management

[6] The Saveso's chemical accident resulted in adoption of the Council Directive 82/501/EEC of June 241,982 on the Major-Accidents Hazards of Certain Industrial Activities, the so-called Saveso directive.

[7] For a holistic perspective on the EU crisis management and analysis of the EU capabilities across policy sectors, institutions and agencies, see: Boin et al. 2013.

[8] Resolution of the Council and Representatives of the Governments of the Member States of 25 June 1987 on the introduction of Community co-operation on civil protection, and subsequent decisions: Resolution of 13 February 1989 on the new developments in Community co-operation on civil protection, Resolution of 23 November 1990 on Community co-operation on civil protection, Resolution of 23 November 1990 on improving mutual aid between Member States in the event of a natural or man-made disaster, Resolution of 8 July 1991 on improving mutual aid in the event of a natural or technological disaster.

[9] Council Decision 2001/792/EC of 23 October 2001 establishing a Community mechanism to facilitate reinforced cooperation in civil protection assistance interventions, OJ L 297, 15 November 2001.

[10] Council Regulation (Euratom) 3954/87 laying down maximum permitted levels of radioactive contamination of foodstuffs and of feeding stuffs following a nuclear accident or any other case of radiological emergency, amended by Regulation No. 2218/89 of 18 July 1989.

tools in this field[11]; and the conflicts in former Yugoslavia had a strong impact on the establishment of the EU Common Security and Defence Policy (CSDP), including external crisis management capabilities (Boin et al. 2013).

The EU gradually developed a large number of cooperation tools and procedures in various policy areas in order to support Member States joint response to crises and disasters. However, a common EU's response to large scale or complex crises has been hampered by institutional complexity. First, at the EU level, the coordination tools were sector-specific and divided by institutional responsibilities, while the new types of crises, often *transboundary* in nature, required a holistic (cross-sectoral and cross-institutional) type of response. Second, a joint response to crises was complicated by the fact that each member state had a distinct system and set of rules at national level (Boin et al. 2013).

7.3 The Origins of the EU Integrated Crisis Response – The IPCR's Predecessor

The need to build a more comprehensive, better coordinated and more efficient crisis response capacity at the Union level became even more evident after the 9/11 terrorist attacks in the US. The scale and complexity of the response to the 9/11 attacks made it clear to the European leaders that the Union was lacking crisis management procedures for cross-sector coordination on the strategic and high political level (Larsson et al. 2009). The process to create such an arrangement started in 2004. The Hague Programme called for the establishment of an integrated and coordinated EU arrangement for crisis management within the Union by 1 July 2006.[12] At that point, in addition to the global challenges outlined in the European Security Strategy (ESS) 2003, internal security – with particular reference to possible major internal crises within the Union, with cross-border effects affecting European citizens, vital infrastructure, public order and security – became an essential element of the EU strategic security thinking.[13] In July 2005, soon after the London terrorist attacks, a declaration by the Justice and Home Affairs Council called for the development of '*arrangements to share information, ensure coordination and enable collective decision-making in emergency, particularly for terrorist attacks on more than one Member State*'.[14] As a consequence of terrorist attacks in the US (2001),

[11] The European Food Safety Authority (EFSA) was set up in January 2002 as an independent source of scientific advice and communication on risks associated with the food chain, European Regulation 178/2002.

[12] *The Hague Programme: Strengthening Freedom, Security and Justice in the European Union* approved by the European Council at the meeting on 5 November 2004, point 2.4. Management of crises within the European Union with cross-border effects, OJ C 53/1, 3 March 2005.

[13] *ibid.*

[14] Council Declaration on the EU response to London bombings, Council of the European Union document 11158/05, 13 July 2005, point 7. http://www.consilium.europa.eu/register/en/content/out/?&typ=ENTRY&i=ADV&DOC_ID=ST-11158-2005-INIT. Accessed 3 October 2018.

Spain (2004) and the UK (2005), and in the aftermath of natural disasters such as the Indian Ocean Tsunami (2004) and Hurricane Katrina (2005), the need for improved emergency responses and specific arrangements for strategic crisis management on a high political level have become ever more urgent.

The EU Emergency and Crisis Coordination Arrangements (CCA), a predecessor of the Integrated Political Crisis Response (IPCR), were formally agreed in December 2005 and officially put into use in July 2006.[15] The CCA were established in order to address the need to conduct rapid policy consultations and coordination at EU political level in major crisis situations. Initially, CCA were designed to bring together the Council and the European Commission, as well as Member States' permanent representatives in Brussels. However, the arrangements have never been fully activated. A limited activation – in info-sharing mode – took place on three occasions. In 2008, during the Mumbai terrorist attacks the General Secretariat of the Council (the EU Situation Centre – SITCEN) in agreement with French Presidency activated the CCA Alert Mode for the first time. The CCA webpage was also used as a hub for information exchange between the EU institutions and Member States during the Haiti earthquake and the volcanic ash cloud crisis in 2010.

The main impediment for a full activation of the CCA turned out to be the formulation of definitions of the type of crises that could trigger a response at the EU level. The CCA arrangements were to be used to assist Member States during emergencies that: *have a direct effect on a number of Member States or which would engage the entire Union; or affect more than one Member State simultaneously, or where the interest of several Member States are engaged together with the responsibilities of EU institutions.*[16] Some examples of such crises included: *multiple, co-ordinated conventional terrorist attacks in several Member States; the loss of key trans-EU infrastructure such as telecommunications networks, oil pipelines or air traffic control systems; a major health emergency, such as a flu pandemic, or a serious accident involving a nuclear reactor in a third country or chemical, biological, radiological or nuclear (CBRN) terrorism, whether in the EU or in a third country.*[17] Since the requirements for triggering the CCA were very strict, basically limited to events of catastrophic dimensions, their activation was restricted to extremely severe crises. Over time, it became apparent that the arrangements could benefit from greater flexibility and scalability in order for the EU to be able to address a broader

[15] Council of the European Union document on EU emergency and crisis co-ordination arrangements, 15106/05, 29 November 2005. http://www.consilium.europa.eu/register/en/content/out/?&typ=ENTRY&i=ADV&DOC_ID=ST-15106-2005-INIT. Accessed 3 October 2018.

Justice and Home Affairs Council Conclusions on the EU Emergency and Crisis Response Capacities, Council of the European Union document 9409/06, 1–2 June 2006. http://www.consilium.europa.eu/register/en/content/out/?&typ=ENTRY&i=ADV&DOC_ID=ST-9409-2006-INIT. Accessed 3 October 2018.

[16] Council of the European Union Document 15106/05, 29 November 2005, point 5. http://www.consilium.europa.eu/register/en/content/out/?&typ=ENTRY&i=ADV&DOC_ID=ST-15106-2005-INIT. Accessed 3 October 2018.

[17] *Ibid*, point 6.

range of crises. These observations were further reinforced by Member States evaluations of CCA exercises. Between 2006 and 2010, five annual CCA exercises were organised to test and improve the arrangements.[18] In spite of that, the results of the exercises in real life situations have shown also positive sides of the arrangements, the evaluation reports indicate for instance that the CCA webpage was considered useful by the Member States and the EU Agencies involved in the annual exercises for information sharing purposes. In 2009, the Swedish Presidency launched a first CCA review process, which resulted in more flexible arrangements with the introduction of the concept of *escalation and adequate crisis modes*.[19] At the end of this initial review process, the Member States also agreed that a more in-depth analysis of the CCA was needed, particularly to consider the post-Lisbon Treaty context.

The main problem with the CCA at the time of the Lisbon Treaty entry into force (1 December 2009) was that the use of CCA arrangements seemed too complex to the Member States. As a result, the EU Member States preferred to use ordinary Council proceedings to handle political coordination at the EU level in the event of a crisis. Moreover, the CCA did not take fully into account the new institutional set-up of the EU, such as changes related to the establishment of the European External Action Service (EEAS), as well as the resulting distribution of responsibilities between the General Secretariat of the Council (GSC) and the EU Situation Centre (SITCEN).[20] Nevertheless, there was still a consensus among the Member States on the need for policy coordination and response at the EU level in the event of major crises, especially on the need to improve and reinforce integrated situational awareness, which formed a support *machinery* of the CCA.

In December 2010, the Belgian Presidency proposed to create a 'Friends of the Presidency' (FoP) group tasked with further review of the CCA. The mandate of this body was formulated as follow: '*Taking into account lessons learned from CCA exercises and real-life crises as well as the post-Lisbon legal and institutional framework: examine whether CCA in its current configuration is the politically and strategically agile tool required by the EU as a whole to respond quickly and adequately to a serious crisis situation. As appropriate, develop proposals that incorporate all relevant EU level actors and crisis coordination tools*'.[21] The FoP meetings were convened and chaired by the rotating Presidencies of the Council,

[18] CCA exercise 2010 (CCAEX10) Draft Evaluation Report, Council of the European Union Document 15529/10, 28 October 2010. http://www.consilium.europa.eu/register/en/content/out/?&typ=ENTRY&i=ADV&DOC_ID=ST-15529-2010-INIT. Accessed 3 October 2018.

[19] CCA Standard Operating Procedures, Council of the European Union Document 11949/2/10 REV2, 20 July 2010. http://www.consilium.europa.eu/register/en/content/out/?&typ=ENTRY&i=ADV&DOC_ID=ST-11949-2010-REV-2. Accessed 3 October 2018.

[20] Council Decision 2010/427/EU of 26 July 2010 establishing the organisation and functioning of the European External Action Service, OJ L 201/30, 3 August 2010.

[21] Review of the Emergency and Crisis Coordination Arrangements (CCA): Mandate of the Friends of the Presidency Group, Council of the European Union Document 17308/10, 3 December 2010. http://www.consilium.europa.eu/register/en/content/out/?&typ=ENTRY&i=ADV&DOC_ID=ST-17308-2010-INIT. Accessed 5 October 2018.

and meant to allow for an inter-institutional consultation process by including representatives of the Member States, Commission, and relevant Council bodies.

In the meantime, many aspects regarding implementation of the newly introduced solidarity clause needed to be clarified.[22] The Commission and the High Representative of the Union for Foreign Affairs and Security Policy (HR) were designated by the Treaty to present a joint proposal for a Council decision in this matter and such a proposal was submitted to the Council in December 2012.[23] After the submission of the joint proposal, further consultations based on this document took place within the FoP group, which was initially set up to deal with the CCA revision.[24] At the end, through the consultations within the FoP group the review of the CCA was incorporated into an overarching discussion about the entirety of the EU crisis management tools, and the distribution of responsibilities among the key actors.

The double mandate of the FoP group functioned well: the revision of the CCA was accomplished in June 2013 with the approval by the Council of the new arrangements, renamed the *EU Integrated Political Crisis Response* (IPCR),[25] and 1 year later the consultations on the joint proposal related to the solidarity clause were concluded. The long consultation process ended in June 2014 with the adoption by the Council of the Decision on the rules and procedures for the implementation by the Union of the solidarity clause.[26]

As mentioned in the introduction, it is important to emphasize that the decision regarding the solidarity clause regulates the implementation of Article 222 of TFEU *by the Union*. While its objective is to ensure coherence and complementarity between the EU and Member States' actions, the Council's decision does not

[22] Article 222(3) TFEU stipulates that: *The arrangements for the implementation by the Union of the solidarity clause shall be defined by a decision adopted by the Council acting on a joint proposal by the Commission and the High Representative of the Union for Foreign Affairs and Security Policy.*

[23] Joint Proposal for a Council Decision on the arrangements for the implementation by the Union of the Solidarity Clause, JOIN (2012) 39 final, 21 December 2012.

[24] COREPER endorsed an additional mandate to the Friends of the Presidency (FoP/CCA) group on the review of the EU Emergency and Crisis Coordination Arrangements to include also the proposal on the implementation of the solidarity clause, thus changing the name of the working group from FoP/CCA to FoP on the CCA review and the solidarity clause implementation (FoP CCA/SCI). Through this mandate the FoP working group was tasked to bring work forward on the solidarity clause and ensure coherent handling of the consultation process in the Council. For details see: Council of the European Union Document 6598/13, 23 February 2013. http://www.consilium.europa.eu/register/en/content/out/?&typ=ENTRY&i=ADV&DOC_ID=ST-6598-2013-INIT. Accessed 5 October 2018.

[25] Finalisation of the CCA review process, Council of the European Union Document 10708/13, 7 June 2013; the Annex on the EU Integrated Political Crisis Response (IPCR) arrangements was approved by the General Affairs Council at the meeting on 25 June 2013. http://www.consilium.europa.eu/register/en/content/out/?&typ=ENTRY&i=ADV&DOC_ID=ST-10708-2013-INIT. Accessed 5 October 2018.

[26] Council Decision 2014/415/EU of 24 June 2014 on the arrangements for the implementation by the Union of the solidarity clause, OJ L 192/56, 1 July 2014.

regulate the implementation by Member States of the solidarity clause within the meaning of Article 222(2) TFEU. According to paragraph 2, and Declaration 37 on Article 222 of TFEU: *a Member State can choose the most appropriate means to comply with its own solidarity obligation towards another Member State.*[27] Consequently, the Member States retain their own competence to decide on the means and extent of assistance that they are willing to provide to a Member State overwhelmed by a crisis, but at the same time they are obliged to coordinate their efforts between themselves in the Council in order to comply with their solidarity obligation. The new decision thus primarily provides rules and procedures for coordination of a response at the EU political level. In addition, according to the principle of subsidiarity, the solidarity clause (and therefore also the IPCR) cannot infringe on Member States responsibilities: *'Since the objective of this decision, namely implementation by the Union of the Solidarity Clause, cannot be sufficiently achieved by the Member States but can rather, by reason of the scale and effects of the action, be better achieved at the Union level, the Union may adopt the measures, in accordance with the principle of subsidiarity as set out in Art.5 of the TEU.'*[28]

7.4 New Crisis Definitions

The scope and activation procedure of the new IPCR arrangements need to be examined in the light of two above-mentioned Council decisions. The reason for this is that the IPCR can be activated by the Presidency of the Council either independently of the solidarity clause invocation,[29] or immediately once the solidarity clause has been invoked by a Member State overwhelmed by a crisis.[30] The IPCR arrangements define the types of crisis in which the IPCR can be activated (independently of the solidarity clause) in a much broader fashion than the previous CCA. The IPCR refers to crisis and crisis situations such as: '*major emergencies or crises, whether inside or outside the EU, of such a wide-ranging impact or political significance, that they require timely policy coordination and response at EU political level. This could result from a number of affected or involved Member States, or the cross-sectoral nature of the crises, the imminence thereof, or from time constraints, or combination of these factors*'.[31]

---

[27] Declaration 37 on Article 222 of the Treaty on the Functioning of the European Union, OJ 2012/C 326/01, 26 October 2012.

[28] Paragraph 20, Council Decision 2014/415/EU, *op. cit.*

[29] Council of the European Union, Document 107008/13, *op. cit.*

[30] Article 5(1), Council Decision 2014/415/EU, *op. cit.*

[31] Paragraph 1 of the Annex, Council of the European Union Document 10708/13, references to 'crisis' or crisis situations' throughout this document cover crises as indicated in this paragraph. http://www.consilium.europa.eu/register/en/content/out/?&typ=ENTRY&i=ADV&DOC_ID=ST-10708-2013-INIT. Accessed 8 October 2018.

In case of the solidarity clause, Article 222 of TFEU refers to three broad threats: terrorist attacks, natural disasters and man-made disasters. Its scope is therefore very broad, and it touches upon a wide array of existing policies. The Council Decision on the implementation of the solidarity clause defines the above mentioned types of crisis in the following terms: '*Disaster* – means any situation which has or may have a severe impact on people, the environment or property, including cultural heritage; *Terrorist attack* – means a terrorist offence as defined in the Council Framework Decision 2002/475/JHA of 13 June 2002 on combating terrorism[32]; *Crisis* – means a disaster or terrorist attack of such wide-ranging impact or political significance that it requires timely policy coordination and response at Union political level.'[33] The inclusion of such broadly defined crises indicates a flexible approach and fairly wide perspective on potential threats. The solidarity clause therefore adopts an all-hazards approach, wider than any specific policy field and means of assistance (Nimark 2015).

7.5 The IPCR – Coordination and Decision-Making

The IPCR is not a new instrument but rather a new set of arrangements (rules, procedures, support tools) that came out of the revision of the CCA. Its analysis requires that it be put in the context of its predecessor. Even though the lack of full activation of the previous arrangements could be considered as their failure, we need to acknowledge that the CCA contributed (in particular through annual exercises 2006–2010) to the development of a more dynamic and integrated culture of crisis coordination and response. Improvement of the inter-institutional cooperation was indeed one of the main aims of the CCA inception. Since 2006, the Council together with the European Commission and later the EEAS, have developed more robust and efficient procedures to respond to emergencies and crises at the EU level. In many respects these have in fact reduced the need for the specific procedures initially contained in the CCA. Consequently, the solidarity clause and the IPCR were oriented at fully exploiting synergies between stakeholders and existing means, structures and capabilities at the EU level.[34] The idea that the IPCR contributes to the implementation of the solidarity clause also reflects the principle of a single set of arrangements and an overall simplification of procedures. The main purpose of this solution was to avoid duplication of existing structures. The IPCR is based on

[32] Article 1, Council Framework Decision 2002/475/JHA of 13 June 2002 on combating terrorism, OJ L 164, 22 June 2002.

[33] Article 3, Council Decision 2014/415/EU, *op.cit.*

[34] Council Decision 2014/415/EU *op. cit.*, Paragraph 4 states that: *The implementation of the solidarity clause by the Union should rely on existing instruments to the extend possible, should increase effectiveness by enhancing coordination and avoiding duplication, should function on the basis of no additional resources, should provide a simple and clear interface at Union level to Member States, and should respect the competences conferred upon each Union institution and service.*

existing procedures and working groups/committees (no ad-hoc bodies), implying that Council working parties and Committee of Permanent Representatives (COREPER) can be convened in a timely manner. Drawing on the well-known, regular Council procedures instead of having recourse to pre-defined ad-hoc groups is intended to allow for swift and efficient action during a crisis. The main aim of the IPCR is simply to reinforce the EU's ability to take rapid decisions when faced by major emergencies.

7.6 A New Name – A New Approach

The change of name from *Emergency and Crisis Coordination Arrangements* to *Integrated Political Crisis Response* arrangements reflects clearly the approach adopted in crisis response. Decision-making at the strategic level in response to a crisis is a *political* process, decisions are to be taken by COREPER and/or the Council (including if needed the European Council) and the Member States through their representatives will have full control over that process. While the IPCR strengthens the political process, it does not replace any sector-specific mechanisms. It is primarily a political coordination arrangement without any additional resources linked to it. The rotating Presidency of the Council (which as a rule chairs the COREPER and other committees' meetings) has a central role to play in order to ensure political control and strategic direction of the IPCR use. In fulfilling its role, the Presidency will also consult the Member States. We can therefore conclude that, so far, the Member States are not willing to delegate their authority of decision making to the EU Institutions and the process remains intergovernmental. The *integrated* aspect of the response means that the Presidency of the Council will be advised and supported by an *informal round table* bringing together all relevant stakeholders (the General Secretariat of the Council, the Commission, the EEAS) in order to consider all relevant means of response available to the EU. The Presidency may also seek expertise from the relevant EU Agencies or Member States' experts (Nimark and Pawlak 2014).

7.7 Improved Flexibility and Scalability

The new arrangements are designed to be more flexible and scalable than the CCA, allowing a tailored involvement at the political level and the required support in relation to a particular crisis. Unlike the CCA, there is no threshold for the IPCR activation and no ad-hoc bodies (such as the CCA *Crisis Steering Group*) involved. From information sharing, analysis and awareness to coordination and decision-making, the arrangements can be adjusted according to the needs of a particular crisis situation. The IPCR is organised around the COREPER institutional set-up, relying on regular, well-known and tested Council procedures. The role of

COREPER is not random; it stems from its cross-sectoral responsibilities, decision-making powers as well as the possibility to convene quickly in Brussels. Depending of the character of the crisis, its consequences, and the related political needs, the Presidency will decide, on the appropriateness of convening a meeting of the relevant Council working parties (e.g. Political and Security Committee – PSC, Standing Committee on operational cooperation on internal security – COSI)[35] and/or COREPER. According to the IPCR procedures, any Member State can request the Presidency to do so and as a consequence to activate the IPCR. The Presidency, or the GSC, Commission and EEAS in full agreement and associating the Presidency, can also decide to activate the IPCR in *information sharing* mode for a limited period of time, in order to prepare ground for a possible full activation. This option does not imply full activation *per se*. The scalability of the IPCR process also implies that depending on the extent and severity of a crisis, decisions can be taken at different political levels – from COREPER to European Council if needed (de Miguel Beriain et al. 2015).

7.8 The IPCR Key Support Elements

It is essential for an authority in charge of the response to a crisis to have the capacity to make sense of an evolving threat/crisis; to be able to initiate a timely and effective response and to be able to communicate effectively (Boin et al. 2013). Even though the EU has a large number of early warning systems, information networks, centres and crisis rooms, the capacity to gather, analyse and disseminate critical real time information is still being developed. What is needed is a capacity that can provide an integrated and shared picture of a situation to the decision-makers (Boin et al. 2014a). The CCA experience has shown that the EU didn't have strong enough capacity to collect and analyse data in real time. Therefore, an important part of the CCA review focused on the establishment of better analysis and awareness instruments, which could develop a common picture of a crisis situation and support decision-making. The overall IPCR architecture comprises a number of supporting elements, such as the *Presidency's informal round table* (responsible for preparation, development and update of proposals for action), the other key support elements are the *Integrated Situational Awareness and Analysis (ISAA) and the IPCR web platform*. Their role is to provide information, analyses and services that are tailored to the needs at the political decision-making level, under the guidance of the Presidency.

[35] COSI is a new Standing Committee introduced by the Lisbon Treaty, Article 71 of TFEU. It has been set up within the Council and its main function is to facilitate, promote and strengthen coordination of operational actions of the authorities of the Member States competent in the field of internal security. See: Council Decision 2010/131/EU on setting up the Standing Committee on Operational Cooperation on Internal Security, 25 February 2010. OJ L 52, 3 March 2010.

The ISAA is a new *sense-making* capability that should support the Presidency and Council's decision-making. This capability is a pivotal element underpinning the IPCR and solidarity clause. The ISAA implementation is not completed yet; the Commission and the EEAS are still developing jointly their Standard Operating Procedures (SOPs). An initial concept paper of the ISAA was presented to the FoP group in November 2013, and since then the SOPs have been further developed.[36] The ISAA is based on existing means and it incorporates structures such as the recently established *Emergency Response Coordination Centre*[37] and *EU Situation Room.*[38] It should also use relevant information and analysis provided by the Member States (e.g. from national crisis centres), particularly through the IPCR Web Platform, and by the EU Agencies. Upon activation, ISAA support should be available on a lasting basis. This new capability is meant to provide an integrated overview of the situation, as well as its evolution and consequences (Boin et al. 2014b).

During the revision process of the CCA, there have been several calls for further improvement of the CCA webpage that was considered by the Member States to be a useful tool for crisis coordination. The *IPCR web platform* is an outcome of this revision and another support capability of the IPCR arrangements. The Web Platform is a key IPCR information-sharing tool – it receives contributions from Member States, the Commission, the EEAS and EU Agencies – acting therefore as the IPCR communication hub. The General Secretariat of the Council manages the platform, with support of the EEAS and the Commission within their respective responsibilities. The platform is accessible to all relevant stakeholders at Member States and EU levels, it is permanently available, and its access is password protected. In time of crisis, one or several Crisis Pages can be generated, depending on the situation and political needs.

The IPCR Web Platform also allows information sharing outside times of crisis, notably for preparedness purposes (networking, exchanging information – including on ongoing situations outside an IPCR activation). The platform had been used to monitor complex crisis situations such as the Ebola outbreak, the Central African Republic crisis, the refugee crisis, the conflicts in Yemen, Syria, Iraq and the earthquake in Nepal. Some of the latest improvements in security and functionality

[36] As of October 2015, the ISAA SOPs were still under discussion within the Friends of the Presidency Group, Council of the European Union Document, 4297/15, 21 October 2015. http://www.consilium.europa.eu/register/en/content/out/?&typ=ENTRY&i=ADV&DOC_ID=CM-4297-2015-INIT. Accessed 8 October 2018.

[37] The Emergency Response Coordination Centre (ERCC) was established in May 2013 and is situated in the Commission's DG ECHO; it is an enhanced coordination platform. The ERCC has a 24/7 monitoring capacity that enables an immediate response to emergencies. It also provides channels for real-time coordination and information sharing through videoconferencing, allowing the centre to connect relevant member states authorities and EU institutions. See Decision 1313/2013/EU of the European Parliament and of the Council of 17 December 2013 on a Union Civil Protection Mechanism, OJ L 347/924, 20 December 2013.

[38] The EU Situation Room was established in July 2011 within the EEAS, the formal duties of this capability are: to provide worldwide monitoring and situation awareness, to serve EU Delegations and CSDP missions, to provide support to EEAS Crisis Management Platform and to engage with relevant Member States' crisis coordination centres.

of the platform included: a new Structured Information Collection Tool to support ISAA that has been developed by the GSC in close coordination with the relevant EU crises centres; a new Situation Monitoring Module available on the preparedness corner that should allow IPCR stakeholders to post information and their existing products regarding a complex crisis but outside an IPCR activation; an archive of the monitoring pages; and development of topical *Hubs* allowing the exchange of information not strictly related to specific ongoing crisis situations.[39]

7.9 IPCR Operational Aspects – First Time Full Activation

In October 2015 the European Council had taken a number of decisions in response to the refugee and migration crisis.[40] The scale of migratory flows, in particular through the Western Balkans migration route, clearly demonstrated the need for coordinated information sharing among the EU Member States and institutions. On October the 25th, the Luxembourg EU Presidency decided to trigger the IPCR arrangements in the *information-sharing* mode.[41] The aim of this measure was to monitor the development of migratory flows, to support decision-making and to better implement the agreed measures. Working closely with the General Secretariat of the Council, the Commission, and the EEAS, the Presidency was able to quickly define the political and strategic guidance for the preparation of ISAA reports. This guidance, shared with all Member States via the IPCR web platform, provided the main areas of interest for the collection of information and the timeframe. The ISAA reports started to be delivered on a weekly basis helping to instantly share the overview of the situation among all member states.

On November 9, in reaction to the rapid worsening of the refugee crisis, the Council of the EU decided for the first time to fully trigger the IPCR.[42] The full activation of the IPCR meant that on the basis of needs and gaps identified in the ISAA reports, the Presidency was able to regularly convene roundtables for the

[39] Council of the European Union Document 7051/14, 27 February 2014 and 11417/15. http://www.consilium.europa.eu/register/en/content/out/?&typ=ENTRY&i=ADV&DOC_ID=ST-7051-2014-INIT. Accessed 8 October 2018.

[40] Conclusions of the European Council Meeting, 15 October 2015. http://www.consilium.europa.eu/en/meetings/european-council/2015/10/15-16/. Accessed 10 April 2017.

[41] Council of the European Union Document 6745/16, 3 March 2016, Refugee and Migration Crisis focused analysis of the use of the IPCR Web Platform. http://data.consilium.europa.eu/doc/document/ST-6745-2016-INIT/en/pdf. Accessed 8 October 2018.

[42] Council of the European Union Document 13880/15, 9 November 2015, Annex: Council Conclusions on Measures to Handle the Refugee and Migration Crisis, Point 18: *to support the Presidency's Decision to upgrade the activation of the IPCR from information-sharing to full activation mode including by providing information to feed the ISAA process, by contributing to the identification of operational gaps, and by supporting communication activities, thereby improving political coordination and decision-making process at the EU level.* http://www.consilium.europa.eu/register/en/content/out/?&typ=ENTRY&i=ADV&DOC_ID=ST-13880-2015-INIT. Accessed 8 October 2008.

management of the migration crisis, bringing together appropriate expertise in order to allow timely policy coordination.[43] The Luxembourg Presidency organised 3 IPCR roundtable meetings at Ambassador level, and 10 in a working-level format chaired by a JHA counsellor. The topics addressed at the meetings included the humanitarian situation, hotpots, security checks, smuggling and trafficking migrants, internal borders, return and readmission. Representatives of the relevant EU agencies (such as Frontex or Europol) and international organisations (like the UNHCR or IOM) were invited by the Presidency to the IPCR meetings in order to share their expertise on particular topics. On some occasions, the representatives of the non-EU states (e.g. Serbia, Former Yugoslav Republic of Macedonia) or NGOs like the ICRC were also invited to the IPCR meetings.[44]

In November 2016, the EU and Turkey held a Summit and adopted a Joint Action Plan to deal with the refugee crisis. The implementation of the actions agreed at the Summit was also supported via the IPCR and was the topic of a roundtable meeting at Ambassadors level within the 28 EU member states. This meeting was followed by a working level roundtable that allowed to launch, together with Turkey, a joint data gathering exercise on migration flows and support refugees in Turkey. The roundtable meetings have become the core of the IPCR mechanism during the most intensive face of the refugee crisis. The subsequent EU rotational presidency, held by the Netherlands in the first half of 2016, organized 32 IPCR roundtable meetings and produced 3 presidency reports on migration.[45] The meetings have continued to be organised by Slovakia and Malta during their Presidencies in the second half of 2016 and the beginning of 2017, although less frequently.

While at the political level there has been a lot of disagreements between the EU member states regarding the admission and relocation of the refugees, activation of the IPCR helped the EU to coordinate its response to the refugee crisis in certain regards. For example, the IPCR arrangements improved the implementation of interlinked measures on borders, reception capacities, hotspots and returns. By using the IPCR the EU was also able to better coordinate financial and human resources pledges as well as operational and logistical priorities.[46] One of the direct outcomes of the discussions taking place within the IPCR roundtables was the adoption of the Council Regulation 2016/369 on the provision of emergency support within the Union.[47] The Netherlands Presidency stressed repeatedly in the

---

[43] For a detailed evaluation of the first IPCR activation see: Annex to the Luxembourg EU Presidency Report – Managing migration flows, State of play – implementing solutions and remaining gaps, 16 December 2015. http://www.eu2015lu.eu/en/actualites/notes-fond/2015/12/17-migration-flows-management-report-presidency/12_17_migration-flows-management-report-presidency.pdf . Accessed 8 October 2018.

[44] *Ibid.*

[45] https://english.eu2016.nl/documents/reports/2016/02/13/presidency-report-migration. Accessed 7 October 2018.

[46] The Netherlands EU Presidency Report, Joining efforts on migration, 7 March 2016. https://english.eu2016.nl/documents/reports/2016/02/13/presidency-report-migration. Accessed 7 October 2018.

[47] Council Regulation (EU) 2016/369 of 15 March 2016 on the provision of emergency support within the Union. http://eur-lex.europa.eu/legal-content/EN/TXT/?uri=CELEX%3A32016R0369. Accessed 7 October 2018.

roundtables meetings of the IPCR the need for humanitarian aid to become available within the EU in order to reduce the economic impact of the refugee crisis in countries such as Greece or Italy. The European Commission prepared a relevant proposal for a Council Decision that would allow application of humanitarian aid within the Union and the Regulation was adopted by the Council in March 2016.

The cross-sectoral and cross-border character of the refugee crisis provides a good testing ground for the IPCR (Minard 2015). The state of play of the IPCR activation, ISAA information-sharing and initial lessons learned are being continuously discussed within the Friends of the Presidency Group and will be integrated in the overall IPCR evaluation process once the crisis is over.[48] While the IPCR remains activated as of September 2018, a first external report evaluating the EU response to the migration crisis was published by the Migration Policy Institute in June 2018 (Collett and Le Coz 2018).[49] The report offers a first reflection on the formal and informal crisis-response structures that evolved to manage the arrival of migrants and asylum seekers on the EU territory between 2014 and 2017. The authors of the report note in the introduction that given the fast-paced nature of both crisis response and institutional reforms, some of the mechanisms are still evolving. Therefore, the impact of the unfolding policy and operational developments in crisis management is still unclear, as is their future status. Referring specifically to the role of the IPCR arrangements, the report states in its key recommendations that: "*The European Union cannot and should not, remain in permanent crisis mode. Currently, EU institutions are wary of deactivating key crisis coordination mechanisms for fear of losing the ability to quickly react to change. The European Union needs to set in place a series of non-crisis mechanisms that can both flag concerns effectively and escalate responses when needed*". The report recognises the central role of the IPCR during the crisis and proposes that the key elements of the new arrangements should be translated into an Integrated Political Migration Response (IPMR) to allow the critical reporting of the mechanism to continue in non-crisis mode.

7.10 Conclusion

The adoption of new mechanisms aimed at simplifying and consolidating the EU's response to major crises is a recent development in the construction of the EU crisis management. Since the migration crisis is still ongoing and the IPCR arrangements remain activated, it is difficult at this stage to evaluate fully their effectiveness. With their first activation, the IPCR arrangements have already gone further than their predecessor, the CCA. The IPCR SOPs (and support measures such as the ISAA and the Web Platform) had been developed by 2015 just in time for the first IPCR

[48] Council of the European Union Document, CM 4452/18, 24 September 2018. http://data.consilium.europa.eu/doc/document/CM-4452-2018-INIT/en/pdf. Accessed 8 October 2018.

[49] An earlier version of the report was commissioned by the General Secretariat of the Council to inform internal discussions with senior EU and national officials.

activation but they remain living documents. Therefore, lessons learned at the end of the current activation will be of great importance and should help the EU policy-makers to build sustainable mechanisms to manage future emergencies.

We can draw some initial conclusions about the IPCR based on the analysis of the consultation process that brought this new instrument into life. Considering that the establishment of the IPCR arrangements had resulted from a thorough revision of the CCA arrangements that took place between 2010 and 2013, lessons learned from the CCA helped to design an improved set of procedures and support tools. Another positive aspect of the CCA revision that needs to be highlighted is the inter-institutional way of conducting the consultation process. The work done by the FoP group was based on the consultation between the Member States, the General Secretariat of the Council, the EEAS and the Commission, and helped to build synergy and consensus among all relevant stakeholders. Moreover, the double mandate of the FoP for the CCA/IPCR and the solidarity clause implementation permitted the adoption of an integrated approach to crisis responses. One set of arrangements in the form of the IPCR, which can be activated in all major crisis situations, brings a promising simplification of the decision-making procedures.

Similarly, to the CCA, the IPCR regulates how the EU Member States and institutions can cooperate on a strategic and high political level in the event of a major crisis. The role of the Member States in the Post-Lisbon institutional set-up remains without change; the Member States provide a sort of *coordinative authority* to the EU in times of crises but do not cede their sovereign rights and primary responsibilities in national security area (Boin et al. 2013). The central role of the Council (including the role of COREPER and rotating Presidency) highlights the political character of the decision-making process. The use of the COREPER's well-known procedures should definitely help to increase the recognition of the EU authority for decision-making in complex crisis situations.

The implementation of the IPCR arrangements brought some clear improvements. The EU has now at its disposal a generic crisis management mechanism that can be used in crisis situations originating inside or outside the EU as long as the crisis affects one or a number of EU Member States. Due to clarification of the definitions of crises and disasters, the new arrangements are more flexible and scalable than the CCA; there are no more thresholds regarding the triggering of the IPCR and the level of political response can be adjusted accordingly to the needs of a particular crisis. Previously, the EU did not have a strong capability to collect and analyse data in real time, but the establishment of the new support measures such as the ISAA and the IPCR web platform should strengthen this aspect of crisis response. Most importantly, the IPCR as a new coordinative arrangement should help to overcome institutional complexity and difficulties of coordination within and between the EU institutions as well as between the EU and its Member States.

References

Boin, Arjen, Magnus Ekengren, and Mark Rhinard. 2013. *The European Union as crisis manager: Patterns and prospects*. Cambridge: Cambridge University Press.

Boin, Arjen, Ekengren, Magnus, and Rhinard, Mark. 2014a. *Making Sense of Sense-Making: The EU's Role in Collecting, Analysing and Disseminating Information in Times of Crisis*. http://www.societalsecurity.eu/uploads/Articles/2014_Boin%20Ekengren%20Rhinard_Sensemaking_FHS%20Book.pdf. Accessed 7 Oct 2018.

Boin, Arjen, Magnus Ekengren, and Mark Rhinard. 2014b. Sensemaking in crises: What role for the EU? In *Crisis rooms – Towards a global network?* ed. Patryk Pawlak and Andrea Ricci, 117–128. EU Institute for Security Studies. https://www.iss.europa.eu/content/crisis-rooms-—towards-global-network. Accessed 7 Oct 2018.

Collett, Elizabeth and Le Coz, Camille. 2018. *After the storm. Learning from the EU response to the migration crisis*. Migration Policy Institute. https://www.migrationpolicy.org/research/after-storm-eu-response-migration-crisis. Accessed 8 Oct 2018.

De Miguel Beriain, Inigo, Atienza-Macias Elena, and Armaza Armaza Emilio. 2015. The European Union integrated political crisis response arrangements: Improving the European Union's major crisis response coordination capacities. *Disaster Medicine and Public Health Preparedness* 9 (3): 234–238.

Larsson, Per, et al. 2009. The Crisis Coordination Arrangements (CCA). In *Crisis management in the European Union: Cooperation in the face of emergencies*, ed. Olsson Stefan, 127–138. Berlin/Heidelberg: Springer.

Minard, Pierre. 2015. *The IPCR arrangements: a joined-up approach in crisis response?* EU Institute for Security Studies. https://www.iss.europa.eu/content/ipcr-arrangements-joined-approach-crisis-response. Accessed 7 Oct 2018.

Nimark, Agnieszka, and Patryk Pawlak. 2014. Upgrading the Union's response to crises. In *Crisis rooms – Towards a global network?* ed. Patryk Pawlak and Andrea Ricci, 107–115. EU Institute for Security Studies. https://www.iss.europa.eu/content/crisis-rooms-—towards-global-network. Accessed 7 Oct 2018.

Nimark, Agnieszka. 2015. The EU's Role in Crisis and Disaster Management within the Union: Post-Lisbon Developments. In *Essays on Global Safety Governance: Challenges and Solutions*, ed. Patrycja Dabrowska-Klosinska, 149–172. Warsaw: Centre for Europe, University of Warsaw.

Chapter 8
The Solidarity Clause: A Review

Iñigo de Miguel Beriain

Abstract The Solidarity Clause constitutes one of the most relevant tools of the European Union (EU) legal framework on major crises management. Created by the Treaty of Lisbon, the Clause is considered to be a nice example of the advantages provided by cooperation among the EU and its Member States (MS) in so far as it is meant to guarantee an adequate division of the tasks and duties between both types of agents. However, its real utility is yet to be checked, since it has never been invoked. Therefore, some important issues related to the Clause remain still unclear, such as the concrete sense of the term "solidarity", its relationship with the Mutual Defence Clause, the level of intensity that a crisis should reach to allow activation of the mechanism, the consequences of a violation of the Clause, etc. This chapter is intended to bring some light to these challenging issues through a careful analysis of the Solidarity Clause.

Keywords Solidarity clause · Mutual Defence clause · Crisis preparedness and response · EU cooperation · Solidarity · EU mutual aid

8.1 Introduction

The Solidarity Clause constitutes one of the most relevant tools of the European Union (EU) legal framework on major crises management. Created by the Treaty of Lisbon, the Clause is considered to be a nice example of the advantages provided by cooperation among the EU and its Member States (MS) in so far as it is meant to guarantee an adequate division of the tasks and duties between both types of agents. As Roderick Parkes has stated, the Clause creates a "sort of division of

I. de Miguel Beriain (✉)
University of the Basque Country, UPV/EHU,
Leioa, Spain

IKERBASQUE, Basque Foundation for Science, Bilbao, Spain
e-mail: INIGO.DEMIGUELB@ehu.eus

D. P. O'Mathúna, I. de Miguel Beriain (eds.), *Ethics and Law for Chemical, Biological, Radiological, Nuclear & Explosive Crises*, The International Library of Ethics, Law and Technology 20, https://doi.org/10.1007/978-3-030-11977-5_8

labour between the EU and the Member States (…), with the EU trying to be the brain (i.e. do the analysis, pull together the resources) and the MS being the arm (with their resources and political legitimacy)" (Bonacquiti 2015: 12). Therefore, we have good reasons to welcome its existence, especially in times when the probability of the occurrence of a man-made disaster seems to be higher than ever.

However, it is also necessary to highlight that the Clause is yet a quite unknown tool, whose real utility has never been tested in practise, as the Clause has never been activated. Moreover, doubts still remain about some crucial questions related to its scope, its binding force, the mechanisms of implementation, etc., which are in urgent need of answers. This paper describes the Clause in legal terms and also tries to expose all these issues, while providing some suggestions on how to deal with them so that the Clause might help to "provide the impetus for enhancing the EU's leverage among European citizens, offering tangible evidence of the benefits of increased EU cooperation in terms of crisis management and disaster response capabilities" (EU Parliament 2012). It provides a complementary analysis to Chap. 9, which although also addressing solidarity, examines the concept from a philosophical and ethical perspective.

8.2 Historical Background

The roots of the Solidarity Clause must be traced to the terrorist attacks that took place in the USA on 11 September 2001. Indeed, this tragedy must be described as a decisive milestone in the construction of the EU's common management and response to major crisis situations. It was very soon after this that both the European Union and its Member States realized that it was absolutely necessary to strengthen their cooperation tools to adequately face this new kind of threat. As the Commission stated, "The need for a Europe-wide approach to civil protection took on a new sense of urgency after the terrorist attacks in the USA on 11 September 2001. Member States quickly realized that the Union would need a clear, coordinated disaster-response strategy if a similar attack were to take place in a Member State" (European Commission 2002: 4). One of the major results of this new spirit was undoubtedly the creation of the so-called "Solidarity Clause".

The Solidarity Clause was firstly formulated during the European Convention debates on a draft constitution for the EU (2000–2003). At that time, Working Group VIII, related to defence issues, realized that an adequate response to the new threats posed by terrorist attacks needed a broader focus than the traditional common military response expressed by the "mutual defence clause". As a result, they adhered to the idea, primarily expressed by the chair of the group, the former European Commissioner and French Minister Michel Barnier, to support a stronger role for the EU not only in intentional threats but also in managing unintentional disasters, both man-made and natural (Myrdal and Rhinard 2010). This new approach, which was clearly strengthened after the March 2004 terror attacks in Madrid, had a decisive influence on the final redaction of the clause (Blockmans 2014).

As commonly known, the Constitutional Treaty never came into force, due to the rejection of two major States, France and the Netherlands. The Solidarity Clause, however, was finally incorporated into the Treaty of Lisbon (signed at Lisbon, 13 December 2007), which was developed as an alternative option to the failed Constitution. Therefore, the clause became part of the EU core legal framework, included in Article 222, which states as follows:

> *1. The Union and its Member States shall act jointly in a spirit of solidarity if a Member State is the object of a terrorist attack or the victim of a natural or man-made disaster. The Union shall mobilise all the instruments at its disposal, including the military resources made available by the Member States, to:*
> *(a) — prevent the terrorist threat in the territory of the Member States;*
> *— protect democratic institutions and the civilian population from any terrorist attack;*
> *— assist a Member State in its territory, at the request of its political authorities, in the event of a terrorist attack;*
> *(b) assist a Member State in its territory, at the request of its political authorities, in the event of a natural or man-made disaster.*
> *2. Should a Member State be the object of a terrorist attack or the victim of a natural or man- made disaster, the other Member States shall assist it at the request of its political authorities. To that end, the Member States shall coordinate between themselves in the Council.*
> *3. The arrangements for the implementation by the Union of the solidarity clause shall be defined by a decision adopted by the Council acting on a joint proposal by the Commission and the High Representative of the Union for Foreign Affairs and Security Policy. The Council shall act in accordance with Article 31(1) of the Treaty on European Union where this decision has defence implications. The European Parliament shall be informed.*
> *For the purposes of this paragraph and without prejudice to Article 240, the Council shall be assisted by the Political and Security Committee with the support of the structures developed in the context of the common security and defence policy and by the Committee referred to in Article 71; the two committees shall, if necessary, submit joint opinions.*
> *4. The European Council shall regularly assess the threats facing the Union in order to enable the Union and its Member States to take effective action.*

8.3 The Solidarity Clause: Some Preliminary Comments

The text of the Clause was clear enough so as to let everybody arrive at some conclusions about the EU's legal framework on crises response. For instance, it was easy to conclude that it involved a duty of the Union and Member States to act jointly in a spirit of solidarity if a Member State were to be the object of a terrorist attack or the victim of a natural or man-made disaster but under the coordination of the EU institutions, in contrast "to the legal arrangements for internal crises, which are exclusively a national competence and crisis management, which is mainly based on voluntary commitment" (Fuchs-Drapier 2011).

It was also clear that Article 222 included an obligation for the Union to "mobilize all the instruments at its disposal, including the military resources made available by the Member States". This means that all institutional tools, facilities, mechanisms, resources, etc. at the disposition of the EU should be brought into action in case a "Member State is the object of a terrorist attack or the victim of a natural or man-made disaster". This made an enormous difference in comparison with the previously existing European crisis management approach, which was characterized by a multitude of fragmented instruments and policies.

Finally, the Clause also settled an explicit obligation for all EU members, to assist a Member State should it be the object of a terrorist attack or the victim of a natural or man-made disaster, including the possibility of mobilizing military resources made available by other Member States.

However, the redaction of Article 222 also involved a number of essential issues that needed further clarification (Fuchs-Drapier 2011; Myrdal and Rhinard 2010). Some of the issues were:

- The concrete sense of the term "solidarity" was not at all clarified by the wording of the Clause.
- The scope of the Clause remained unclear. Was it applicable both to a national crisis and an international crisis?
- The relationship between the Solidarity and the Mutual Defence clauses was unclear. Did they exclude each other, or could they be triggered simultaneously?
- The level of intensity that a crisis should reach to allow activation of the mechanism was uncertain. The identity of the agent that should make a judgement on this issue was not clear (whether it was the affected MS, the EU institutions, the other MS).
- The consequences of a violation of the Clause were not elaborated. What would happen if a Member State refused to provide aid?

All these issues were meant to be addressed by the arrangements for the implementation mentioned in number 3 of Article 222. This obligation was finally fulfilled several years after the enforcement of the clause by a Council Decision of 24 June 2014 on the arrangements for the implementation by the Union of the Solidarity Clause (O.J.L. 192/53, 1 Jul. 2014), which followed a European Parliament resolution of 22 November 2012 on the EU's mutual defence and Solidarity Clauses. This concerned political and operational dimensions, the opinion expressed by Member States and the Joint Proposal from the Commission and the High Representative of the Union for Foreign Affairs and Security Policy to the Council of the European Union. All these documents provide insights into the issues previously mentioned, and the following paragraphs are dedicated to commenting on them all.

8.4 The Real Meaning of "Solidarity"

"Solidarity" is a quite complex concept as it can be interpreted in several different ways. As Myrdal and Rhinard stated, "for some, solidarity is measured by how much support flows to a country in need. For others, solidarity means everyone doing their own 'homework' to avoid the need for assistance in the first place. Still, others believe that solidarity against today's risks and threats is best pursued outside of EU frameworks" (Myrdal and Rhinard 2010). Depending on the meaning accepted, totally different conclusions arise as to its concrete implications. Indeed, if someone thinks of this idea as synonymous to "altruism", as people usually do in common language, he should conclude that the Solidarity Clause involves a duty to assist a Member State affected by a major crisis scenario, even if it involves no expectation of obtaining any kind of compensation for the resources involved (Hilpold 2015b). However, some authors have stated that this is not in any way an adequate understanding of the legal meaning of "solidarity" according to the EU legal framework as a whole. For instance, Peter Hilpold has stated, in the EU context, "solidarity is based on the principle of reciprocity: 'Solidarity expects solidarity'. This *do et des* mechanism underlying solidarity may not be immediately visible but in the long run and in the broader context all forms of aid and assistance are in some form expected to be repaid" (Hilpold 2015a).

Therefore, it must be kept in mind that the notion of solidarity in Article 222 includes a binding obligation on the Member States to do their best to prepare for crisis situations and, when they occur, to face the challenge on their own and to only invoke the Clause if they feel overwhelmed by the consequences. In fact, the European Parliament resolution of 22 November 2012 on the EU's Mutual Defence and Solidarity Clauses for political and operational dimensions (2012/2223(INI)) "stresses the need for Member States to invest in their own security and disaster response capabilities and not to excessively rely on the solidarity of others; and emphasizes the primary responsibility of Member States for civil protection and security in their territory". Similarly, the Council Decision introduced a clarifying point in article 4 by stating that "in the event of a disaster or terrorist attack, the affected Member State may invoke the Solidarity Clause if, after having exploited the possibilities offered by existing means and tools at national and Union level, it considers that the crisis clearly overwhelms the response capabilities available to it".

To sum up, one must consider that the real aim of the signatories of the Convention was to avoid unnecessary efforts from EU partners while providing them with a rational structure able to support their needs in cases when an unpredictable major disaster overwhelms their capacities to provide an adequate response on their own. Thus, one must think about the Solidarity Clause in terms of a kind of mutual insurance tool able to minimize the risks at a common minimal cost but not in any case an altruistic tool created in order to provide aid for those countries that are not able to face the challenges posed by major crisis situations on their own.

8.5 The Relationship with the "Mutual Defence" Clause

The EU Mutual Defence Clause, also called the EU Mutual Assistance Clause, was incorporated into article 42(7) of the Treaty on the Functioning of the EU (TFEU). In 2015, it became especially relevant due to France's decision to request assistance from the other Member States under this basis in the aftermath of the terrorist attacks in Paris. The Mutual Defence Clause shares some characteristics with the Solidarity Clause, in so far as both of them are meant to lead with major threats to EU. Some authors, such as Blockmans and Wessel, have argued that the existence of the Defence Clause in the TFEU should be considered quite peculiar, keeping in mind the existence of the Solidarity Clause (Blockmans and Wessel 2009). However, some other experts, such as Konstandinides have written that this opinion "does not take into account that despite the confusing similarities between the two provisions there are also fundamental differences" (Konstantidinides 2011).

Indeed, it seems more plausible to hold that both clauses are of a completely different nature. The Defence Clause is a typical mutual aid clause included in military alliance treaties, very similar to Article 5 of the Treaty of Washington in the NATO context. Thus, it is meant to be activated in the event of an attack made by a traditional State against an EU Member State. The Solidarity Clause, on the other hand, is meant to be a kind of "mutual aid" tool, which enters into force whenever a major disaster – any kind of major disaster – is foreseen or has already happened. Therefore, it entails a much broader perspective since it addresses all possible major disaster situations, no matter how they are produced. In this way, the EU tries to provide an adequate answer to all kind of threats (Myrdal and Rhinard 2010). But, even more important, the special relevance of the Solidarity clause in the EU context comes from the fact that it sets the difference between a mere military alliance of different countries and what the EU seeks to be, a real supranational entity based on the idea of solidarity, something that the Mutual Defence Clause does not provide at all. Finally, it is important to emphasise that the institutional framework related to both clauses also differs considerably. The Defence Clause does not involve EU structures but is a purely intergovernmental device. Thus, reaction against an attack is defined on the basis of a State-to-State relationship. Article 222, instead, appeals to coordinated action between the Member States and provides the EU with power to mobilize all instruments at its disposal to guarantee an adequate response (Konstantidinides 2011: 20).

To sum up, it must be concluded that both the Solidarity Clause and the Mutual Defence Clause are different tools that should act as complementary political instruments. However, it is also true that, in practice, the relationship between the Solidarity Clause and the Mutual Defence Clause remains quite obscure for a number of reasons. At the present moment, under some circumstances, deciding which clause to invoke can be difficult for the simple reason that it can be very hard to distinguish between armed aggression from another State and a terrorist attack (Myrdal and Rhinard 2010). This issue triggers a very important question. Under what conditions may the military instrument be used and with what concrete limits? (House of Lords 2009) Which sovereignty should they respect? All these issues seem to concern the case of a terrorist attack as this situation involves the use of both civil and military resources. In such a situation, it is almost impossible to exclude that both clauses apply in parallel. As Hilpold states, "In this case it would be important to find an appropriate way for the coordination of the intergovernmental decision making that applies for Article 42.7 and the EU internal decision making foreseen in Article 222 TFEU" (Hilpold 2015a: 217).

Finally, it is also important to keep in mind that the use of military resources, even if perfectly possible under the umbrella of Article 222, might be much more difficult to activate in the case of a terrorist attack than in the case of man-made or natural disasters in so far as, in some EU Member States, such as Germany, it requires a declared state of defence according to the Basic Law and is only possible with a two-thirds parliamentary majority (Fuchs-Drapier 2011: 185).

8.6 The Scope of the Clause

Another extremely important issue regarding Article 222 refers to its concrete scope, that is, its geographical scope, and the type of crisis that might trigger the legal obligations involved in the Solidarity Clause. Regarding the first issue, it is worth mentioning that article 2 of the Council Decision detailed that the Clause applies (a) within the territory of Member States to which the Treaties apply, meaning land area, internal waters, territorial sea and airspace, and (b) when affecting infrastructure (such as off-shore oil and gas installations) situated in the territorial sea, the exclusive economic zone or the continental shelf of a Member State. Therefore, it seems that the application of Article 222 is excluded when it involves actions to be implemented outside the territory of the Union, leaving apart offshore installations. However, if this is the real intention of the Council, then it must be criticized in so far as disasters affecting Member States might happen outside of the EU's territory (let us remember Chernobyl). Indeed, it would be simply absurd to react to this type of crisis by performing a limited, non-decisive action in a Member State area only on the basis that the interpretation of the Clause impedes a more effective response in the area where the incident was originally provoked. Therefore, it is necessary to appeal for a more creative interpretation of the terms expressed by the Clause or, even more simply, to consider that, if the Member States and the EU institutions decide to go further than the strict terms of the Clause, no illegality would be performed.

The same type of concern arises when one focuses on the types of disasters that Article 222 is meant to deal with. At a first glance, the text of the Clause seems quite clear. It was designed to face both terrorist threats and attacks and to help Member States to provide an adequate answer to natural or man-made disasters. However, a deeper inspection of the text raises some issues that are not so easy to solve. First of all, it makes sense to remember that the definition of "terrorism" is rather vague, an issue that will not be solved by applying another legal tool, such as the Council Framework Decision 2002/475/JHA, as the Council Decision indicates (art. 3b). On the other hand, it remains unclear if the invocation of the clause by a Member State would need (or not) a previous political assessment made by the competent EU institutions, both the Commission's Emergency Response Centre (ERC) and the High Representative for Foreign Affairs and Security Policy (Hilpold 2015a, b). At this point, it is worth mentioning that the Council Decision clarified the meaning of the expression "prevent the terrorist threat", originally included in Article 222, by stating that it refers to an attack in general, that is, leaving out the reductionist focus endorsed by the Joint Proposal from the Commission and the High Representative, which only included the idea of "an actual or imminent terrorist attack" (article 3e). Therefore, it must be concluded that the Clause could be invoked even in those cases when resources are needed in the prevention phase.

Last, but not least, it is unclear what the real scope of the expression "natural or man-made disaster" is. Indeed, this has been especially important in the last years. On 27 October 2015, the *Financial Times* published an article reflecting on how Slovenia was considering its option to trigger Article 222 to cope with the migration crisis (at that time, that State had already received 84,000 immigrants) (Barker and Politi 2015). In the end, Slovenia decided not to proceed. Therefore, uncertainty about the applicability of the Clause to a migration crisis and the political and the practical consequences of its invocation remain totally unknown. It seems quite clear that the use of the Solidarity Clause to face a migration crisis was not in the minds of those who created it. In any case, this fact does not directly negate the possibility of its application. It depends on the interpretation given to the term 'disaster'. The Decision defined it as "any situation which has or may have a severe impact on people, the environment or property, including cultural heritage", which clearly includes massive human migration waves in the territory of the EU. Therefore, if one also considers that such situations are "man-made" in so far they have been provoked by a war, then it is undeniable that the Solidarity Clause could be invoked under the current circumstances.

It is, however, much more difficult to foresee the real impact that this might have, keeping in mind the practical obligations involved in its activation and the real possibility of applying on Member States that refuse to accomplish their duties. Even though this is totally right from a legal perspective, it is hard to believe that the same EU institutions and Member States who permitted the dramatic situation in the refugees' improvised camps in the north of Greece will change their attitude only on the basis of the activation of the Clause. However, its invocation would undoubtedly serve as an additional means of pressure on EU institutions and the richest EU countries, which seem to have forgotten some of the fundamental values included in the EU legal framework, including solidarity. In any case, it will be extremely difficult to check the validity of this statement: if the clause was not invoked in the past, it will hardly be invoked in the future (at least in connection with the refugees' crisis).

8.7 The Nature of the Obligations and the Consequences of a Breach

One of the most important issues for the Solidarity Clause refers to the extent of the commitment. Who decides what resources are to be committed? The EU institutions? The Member State triggering the Clause? The Member States that allocate those resources? Moreover, is a Member State really obliged to provide some minimum amount of aid? To what extent? These questions have required answers for some time. Nowadays, however, both the Joint Proposal and the Council Decision have added some crucial information regarding these issues. First of all, it is now fully evident that the Clause does not involve an obligation for the EU institutions to create new tools or add new resources to the ones already existing, even if Article 222.1 states that "the Union shall mobilise all the instruments at its disposal". As the Council Decision states, "The implementation of the Solidarity Clause by the Union should rely on existing instruments to the extent possible, should increase effectiveness by enhancing coordination and avoiding duplication, should function on the basis of no additional resources, should provide a simple and clear interface at Union level to Member States, and should respect the competences conferred upon each Union institution and service" (recital 4). Therefore, an invocation of the Clause would mobilize several EU instruments, such as the European Union Internal Security Strategy, the European Union Civil Protection Mechanism established by Decision No 1313/2013/EU of the European Parliament and the Council (the Union Mechanism), Decision No 1082/2013/EU of the European Parliament and of the Council, and the structures developed in the framework of the Common Security and Defence Policy (CSDP), but not any others.

Regarding the commitments of the Member States, it is absolutely necessary to highlight that the effective resources to be provided depend on the non-affected Member State. This strong opinion is fundamentally sustained in the Declaration (No. 37) of Article 222 of the Treaty on the Functioning of the European Union, which states that "a Member State can choose the most appropriate means to comply with its own solidarity obligation towards another Member State". Therefore, it will always be the non-affected State who decides on the level of the resources involved, usually organized in modules. Moreover, it will keep the command of the modules sent to the affected State, even if local authorities are responsible for the response to the crisis. In practice, this means that it would be extremely difficult to accuse a Member State of a general inobservance of its obligations in terms of the Solidarity Clause if it is not the case that it directly refuses to send any kind of help with no further justification.

However, what would happen even in that extreme case? Would it be possible to sanction the infringement of the Clause? This issue is controversial and the answer remains unclear. In this respect, Fuchs-Drapier wrote that the European Court of Justice (ECJ) "could theoretically be involved if a Member State refuses to assist another or if capacities for ensuring assistance have not been guaranteed when needed. However, in practice, the ECJ has rather limited scope to scrutinize a Member State's compliance with the Solidarity Clause for legal reason, as well as lack of political will" (Fuchs-Drapier 2011). Moreover, it seems clear that, in so far as the ECJ has no jurisdiction with respect to defence implications, the police and military operations carried out under the application of the Solidarity Clause could not be controlled by this institution. But, keeping this all in mind, what is the real binding effect of the Clause? It is really hard to say. Therefore, it seems quite clear that a country unwilling to comply with the obligations coming from an invocation of the Clause would feel pressured more by political reasons than by purely legal concerns.

8.8 The Implementation Procedure

The implementation of Article 222 would always happen in the event of a major crisis situation but not necessarily in its aftermath. As previously mentioned, it might be invoked even prior to the advent of a crisis if the affected State considered that a preventive intervention might impede it or, at least, minimize its consequences. The most important issue to consider in order to judge if the invocation of the Clause is pertinent or not is whether the affected State feels overwhelmed in its capacity after making all possible effort to face the disaster adequately. If this is the case, the affected Member State is perfectly legitimate in triggering Article 222 mechanisms.

There are two ways to appeal to the Solidarity Clause. First of all, the affected State may directly request another Member State (MS) for help, "although these solidarity duties of the 'horizontal level' are, by far, not so pronounced as those of the 'vertical order', those between the Union and single MS" (Hilpold 2015a: 229). This is further confirmed by the Declaration No. 37 of Article 222 TFEU cited above. Under these circumstances, Article 222 states that both countries "shall coordinate between themselves in the Council" (Hilpold 2015a: 229). However, this does not seem a binding condition for Member States. As Hilpold has written, "Also without this provision the MS would be free to coordinate assistance activities in favour of another MS either inside or outside the Council, both on the basis of international law as within the Common Foreign and Security Policy (CFSP)" (Hilpold 2015a: 221).

The second path consists of an appeal to EU intervention. In this case, there is a quite complex formal procedure designed to guarantee a successful activation of the Clause. First of all, the affected Member State must activate the Clause by notifying its intentions to both the Member State holding the rotating Presidency (who will immediately inform the President of the European Council and the President of the European Parliament) and the President of the Commission through the Emergency response and Coordination Centre (ERCC) (Bonacquiti 2015). In the aftermath of the activation, a quite complex EU architecture enters into action. The EU Integrated Political Crisis Response arrangements (IPCR Arrangements, approved by the General Affairs Council (GAC) 25 June 2013, updating the CCA) are automatically activated in so far as they explicitly state that, under this condition, the Presidency cannot avoid their activation, even if he considers it unnecessary (see the chapter in this book on IPCR by Nimark for further information). As an immediate result, the General Secretariat of the Council sends alerts to inform all stakeholders in Brussels and in the Member States through the IPCR web platform. At the same time, the Commission and the High Representative might mobilize a large number of EU institutions (such as relevant structures, which are in the

Commission (DG ECHO, HOME, SANCO, TAXUD, etc.) or in EU decentralized agencies (FRONTEX, ECDC, EUROPOL, EMSA, EFSA, EMA etc.)). The choice of mobilized institutions will depend on the nature and dimension of the threat. The most important of these is the Commission's Emergency Response Coordination Centre (ERCC), which has a 24/7 monitoring capacity that allows instant reactions to emergencies. This body has its own personnel and a sufficient budget to play its essential role in coordinating the operational response and produce joint situation assessment reports if no other EU institution is more adequate to play that role (Nimark and Pawlak 2013).

8.9 The Solidarity Clause: Final Remarks

After all has been said, it might be concluded that the Solidarity Clause will definitively play a prominent role in the response to major crisis situations, but this role will never substitute that assigned to Member States. This obviously means that they could never relax their response tools in the hope that the EU will provide for the necessary resources to face an improbable chemical, biological, radiological, nuclear or explosive (CBRNE) incident. In fact, this is the scenario that the EU institutions have tried to avoid the most. But, leaving this aside, the legal and institutional framework related to the Solidarity Clause is being clarified, and, hopefully, will become the extremely useful legal resource that it is expected to be.

However, it would be unwise to forget that some extremely important issues regarding the Solidarity Clause remain to be addressed. The most important of these are linked to the lack of a real political will to use this tool to its full extent. The increasing reluctance of Member States to activate binding procedures is limiting their real utility. And, in this case, it is really hard to think that legal refinement might substitute for political will (Fuchs-Drapier 2011). The tension between solidarity and sovereignty remains, and it does not seem that the current political trends will solve this issue. Indeed, the real implications for Member States and their resources in a crisis response, and the level of cooperation required with EU institutions, continue to be undefined (Von Ondarza and Parkes 2010). It must be recognized that the Solidarity Clause will never be able to give an answer to the tension between the requirement to build well-equipped EU structures and the Member States' resistance to automatic solidarity commitments. However, it is also necessary to underline that 90% of European citizens want the EU to take a much more relevant role in disaster response (Special Eurobarometer 2009), this kind of political obstacle will be removed eventually. In this sense, one might adhere to the optimistic approach adopted by Boin, Ekengren and Rhinard when they state that "the contrast between enthusiastic political declarations and hesitant implementation has resulted in patchy capacities at the EU level. At the same time, and somewhat curiously, the EU policy process inexorably muddles on and continues to produce new transboundary crisis management activities" (Boin et al. 2013: 133).

References

Barker, Alex, and James Polity. 2015. Migrant Crisis: Slovenia Considers Calling For Eu Military Aid. *The Financial Times*, October 27. http://www.ft.com/intl/cms/s/0/d47445aa-7cc7-11e5-a1fe-567b37f80b64.html#axzz453bxzbnw. Accessed 3 Oct 2018.

Blockmans, S. 2014. L'Union fait la Force: Making the Most of the Solidarity Clause (Art. 222 TFEU). In *EU management of global emergencies*, ed. Inge Govaere and Sara Poli, 111–135. Leiden: Brill Publishers.

Blockmans, S., and R.A. Wessel. 2009. The European Union and crisis management. Will the Lisbon treaty make the EU more effective? *Journal of Conflict & Security Law* 14 (2): 265–308.

Boin, Arjen, Magnus Ekengren, and Mark Rhinard. 2013. *The European Union as crisis manager. Patterns and prospects*. Cambridge: Cambridge University Press.

Bonacquisti, G. 2015. The solidarity clause: one of the most unacknowledged innovations of the Lisbon Treaty. The European Parliament debates its implementation but also its ambiguities. *Blogactiv.*. https://europe-liberte-securite-justice.org/2015/07/27/the-solidarity-clause-one-of-the-most-unacknowledged-innovations-of-the-lisbon-treaty-the-european-parliament-debates-its-implementation-but-also-its-ambiguities/. Accessed 3 Oct 2018.

Consolidated versions of the Treaty on European Union and the Treaty on the Functioning of the European Union.

European Commission. 2002. *EU Focus on Civil Protection*. http://ec.europa.eu/echo/files/civil_protection/civil/pdfdocs/focus_en.pdf. Accessed 3 Oct 2018.

———. Special Eurobarometer: Civil Protection. November 2009. http://ec.europa.eu/public_opinion/archives/ebs/ ebs_328_sum_en.pdf. Accessed 3 October 2018.

European Parliament resolution of 22 November 2012 on the EU's mutual defence and solidarity clauses: Political and operational dimensions (2012/2223(INI)).

Fuchs-Drapier, M. 2011. The European Union's Solidarity Clause in the Event of a Terrorist Attack: Towards Solidarity or Maintaining Sovereignty? *Journal of Contingencies and Crisis Management*, December, 19(4): 184–197.

Hilpold, P. 2015a. Understanding solidarity within EU law: An analysis of the 'islands of solidarity' with particular regard to monetary union. *Yearbook of European Law* 34 (1): 257–285.

Hilpold, Peter. 2015b. Filling a buzzword with life: The implementation of the solidarity clause in article 222 TFEU. *Legal Issues of Economic Integration* 42 (3): 209–232.

House of Lords. Ordered to be printed 3 March 2009 and published 11 March 2009. European Union Committee, Civil Protection and Crisis Management in the European Union, 6th Report of Session 2008–09. https://publications.parliament.uk/pa/ld200809/ldselect/ldeucom/43/43.pdf. Accessed 3 Oct 2018.

Joint Proposal for a Council Decision on the Arrangements for the Implementation by the Union of the Solidarity Clause of 21 December 2012, JOIN(2012) 39 final.

Konstandinides, T. 2011. Civil Protection in Europe and the Lisbon 'solidarity clause': A genuine legal concept or a paper exercise. Uppsala Faculty of Law, Working Paper 2011:3. https://uu.diva-portal.org/smash/get/diva2:413624/FULLTEXT01.pdf. Accessed 3 Oct 2018.

Myrdal, S., and M. Rhinard. 2010. *The European Union's Solidarity Clause: Empty Letter or Effective Tool?* UI Occasional Papers 2/2010. https://www.ui.se/globalassets/ui.se-eng/publications/ui-publications/the-european-unions-solidarity-clause-empty-letter-or-effective-tool-min.pdf. Accessed 3 Oct 2018.

North Atlantic Treaty of 1949 that created NATO (Treaty of Washington).

Nimark, Agnieszka and Patryk Pawlak. 2013. *Upgrading the Union's response to Disasters*. Brief Issue, European Union Institute for Security Studies, n° 45, November. https://www.iss.europa.eu/sites/default/files/EUISSFiles/Brief_45_Crisis_response.pdf. Accessed 3 Oct 2018.

Treaty of Lisbon amending the Treaty on European Union and the Treaty establishing the European Community, signed at Lisbon, 13 December 2007.

Von Ondarza, Nicolai, and Roderick Parkes. 2010. *The EU in the face of disaster*. Implementing The Lisbon Treaty's Solidarity Clause. Swp Comments 2010/C 09, April.. https://www.swp-berlin.org/fileadmin/contents/products/comments/2010C09_orz_pks_ks.pdf. Accessed 3 Oct 2018.

Chapter 9
Comparing Mutuality and Solidarity in Its Application to Disaster Ethics

Chamundeeswari Kuppuswamy

Abstract Often it has been observed that in disaster situations, people (including victims) become altruistic and are very willing to listen, obey and act in a manner that would help bring an end to the situation. In this chapter, linking disaster ethics and human rights, it is argued that this indeed is how it should be, disaster or otherwise, and that we have moral duties to oneself and to others. An individual exhibiting solidarity, comradery and altruism during a disaster is indeed behaving as a reasonable Self, and exercising ethical individualism as per Gewirthian philosophy. It is the duty of the State and society to act as a supportive State and a caring society. In order to do this, we need to be conditioned for ethical rationality before any whiff of disaster arises, i.e. in our day-to-day conduct and decision-making, at a personal, institutional and transnational level. Our ethical resilience during disasters can only be as robust as our rational moral compass during 'peace-time'. This chapter argues that Gewirthian solidarity ethics (GSE) should play a role in European policy and action in order to provide a system that conditions ethical rationality and in order to fulfil human rights. This involves addressing our current understanding of human rights as distinct categories of civil, political, economic, social and cultural rights and to effect a shift towards a more holistic understanding of human rights, whereby the hierarchy of fulfilment does not always prioritise civil and political rights.

Keywords Solidarity · ESC rights · Human rights · Disaster ethics · Resilience · European disaster preparedness · Moral duties

---

C. Kuppuswamy (✉)
Hertfordshire Law School, Hatfield, UK
e-mail: c.kuppuswamy@herts.ac.uk

D. P. O'Mathúna, I. de Miguel Beriain (eds.), *Ethics and Law for Chemical, Biological, Radiological, Nuclear & Explosive Crises*, The International Library of Ethics, Law and Technology 20, https://doi.org/10.1007/978-3-030-11977-5_9

9.1 Introduction

Often it has been observed that in disaster situations, people (including victims) become altruistic and are very willing to listen, obey and act in a manner that would help bring an end to the situation. This observation applies differently to different people at different points in time – before, during or after the tragic event, but nevertheless there is a definite sense of solidarity, comradery and altruism that can be detected amongst those affected, amongst those responding and in some cases the onlookers. That sense underlies legal applications of solidarity, discussed in Chap. 8 and, in Chap. 10, to the application of solidarity to the ethics of clinical research in chemical, biological, radiological, nuclear and explosive (CBRNE) events. In this chapter, it is argued that this indeed is how it should be, disaster or otherwise, and that we have moral duties to oneself and to others. An individual exhibiting solidarity, comradery and altruism during a disaster is indeed behaving as a reasonable Self, and exercising ethical individualism as per Gewirthian philosophy. It is the duty of the State and society to act as a supportive State and a caring society. It is also noted that in order for the right ethical decisions to be taken in response to disasters, we must already be conditioned to do so by embedding ethical rationality in our preparedness for disaster response, both from the individual point of view as well as the institutional perspective, and also be willing to espouse the same approach in the recovery stages. In order to do this, we need to be conditioned for ethical rationality before any whiff of disaster arises, i.e. in our day-to-day conduct and decision-making, at a personal, institutional and transnational level. Our ethical resilience during disasters can only be as robust as our rational moral compass during 'peace-time'. This is the only reasonable expectation we can hold; anything beyond this is a bonus. But perhaps exceptionally, some behaviour during disaster situations can defy conditioning and make super humans of us.

This chapter argues that Gewirthian solidarity ethics (GSE) should play a role in European policy and action in order to provide a system that conditions ethical rationality and in order to fulfil human rights. In the process, this involves addressing our current understanding of human rights as distinct categories of civil, political, economic, social and cultural rights and to effect a shift towards a more holistic understanding of human rights, whereby the hierarchy of fulfilment does not always prioritise civil and political rights.

Gewirth's *Principle of Generic Consistency (PGC)* is a deontological theory and states that we have moral duties to oneself and to others (Gewirth 1978). Gewirthian solidarity is understood as a categorical imperative, one without a choice, i.e. we are to accept that humans have a duty to protect and uphold human dignity, which, as Beyleveld states, is foundational to action (Beyleveld and Brownsword 2001). In this paper, I argue that while there are numerous conceptions or assumptions about solidarity in official policy discourses, a notion of solidarity on which European solidarity needs to be built has to rely on a strong conception of solidarity as laid out by Gewirth and Beyleveld, if we are to ensure that it is in keeping with our obligations under human rights law. While the applicability of human rights law in the aftermath of disasters has not been so widely examined by regional or international human rights bodies (Barber 2008), it is justified to assume that human rights protection is still integral to disaster reponse. Mutuality in Gewirthian ethics is not reciprocal, but one based on 'basic needs' and works hand in hand with a caring society and a supportive state. A moral structure is built around this idea of mutuality wherein human rights is actually a 'community of rights', and where there is a deep concern about the 'freedom' and 'well-being' of all members of society.

9.2 Altruism and Solidarity in Disaster Situations

Widespread altruistic behaviour is observed in survivors of disasters as well as onlookers, both online and *in-situ*. This has been described as the 'compulsion to help' (Griswold 2013), such as when after the 9/11 attacks, people felt an overpowering sense to volunteer. Crowds have been observed as being benevolent during emergencies (Drury et al. 2009). Recent research even argues for an altruistic brain (Pfaff 2015). In a study that looked into the likely responses by different groups to a bio-terrorist attack, it was found that officials would expect deviant behaviour, but individual citizens would mostly behave altruistically and rationally (Fischer III 2000; Donald and Canter 1992).

Solidarity is exhibited in groups and communities in various ways. The early reports that came out of the terrorist bombings in a Belgian airport and underground transport station in March 2016 recount acts of altruism by victims towards other victims and of those who were not injured towards those affected by the incidents.[1] In 2007, during the London bombings, researchers found 'high levels of mutual aid amongst survivors and witnesses of the bombings' (Drury et al. 2009). Selfishness was more common during the daily rush hour on the tube than during the London bombings on the tube, a telling observation about the desensitisation of individuals on a day-to-day basis. Following the November 2015 terrorist shootings in Paris, companies such as Google pitched in by making their internet call-service free of charge for a certain period. The altruistic behaviour of people in donating blood following the Madrid terrorist bombings in 2004, although not needed for such incidents, was nevertheless in response to an urge to help. A side effect of altruism is also observed. Altruism as a hindrance rather than a help is not uncommon in disaster situations, and points to the duty that exists on members of society to take informed decisions about actions. This inadvertent hindrance can be averted by Gewirthian self-interested rational agents, because their exercise of 'moderate altruism' (Gewirth 1998, 88), does not discount the duty to inform oneself.

In the aftermath of the chemical gas attack in Halabja in northern Iraq in 1988, in the absence of any institutional response, the people of the town buried the dead,

[1] http://edition.cnn.com/2016/03/25/europe/brussels-terror-attack-survivors accessed 12th September 2017; http://edition.cnn.com/2016/03/22/europe/belgium-brussels-attacks-witnesses/index.html accessed 12th September 2017.

demonstrating the poignancy of altruism when their own government ordered the disaster (Hart and Clevestig 2009). During the Tokyo sarin gas attack by the group Aum Shinrikyo in 1995, the lack of obvious signs as to the nature of the attack meant that the first responders to victims were their fellow travellers (Pangi 2002). Reports of callous behaviour of passengers towards 'passengers in discomfort' on the trains in Tokyo could be explained by the fact that there was no obvious sign of an attack. Murakami, in his book that recounts interviews of victims and responders, talks about passengers wanting to distance themselves from the situation by changing trains (Murakami 2000). Perhaps this behaviour is at the cusp of the 'daily commuter syndrome' of indifference and the altruistic individual phenomenon that occurs during an obvious disaster.

Policies that respond to disasters in Europe take into account these behaviours and accommodate, indeed count on, altruism and solidarity. Various initiatives such as EU Aid volunteers provide outlets for such altruism in a structured manner. In the 2009 chemical, biological, radiological and nuclear (CBRN) plan adopted by the EU Commission, the principle of solidarity exists as a top level principle. States have the responsibility primarily in protecting against CBRN threats, but EU initiatives are guided by the EU solidarity principle (Council of the European Union 2009). Article 222 of The Treaty on the Functioning of the European Union (TFEU) incorporates the 'spirit of solidarity' in order to bring together the resources of European states to respond to disasters, and is being infused into the CBRN action plan (Konstadinides 2013, 475). In the US, the 2006 National Strategy for Combating Terrorism highlights the need to develop a 'Culture of Preparedness', and it seems natural to assume solidarity as integral to this culture. While the US is politically less enamoured by solidarity, it has embedded solidarity in its culture of preparedness statement – which incorporates 'vocabularies of cultural change' (Bean 2011). This refers to the culture of information sharing between departments as opposed to a culture of need to know. Although not a radical step in solidarity, but a essential first step towards it.

Solidarity amongst niche victim groups, as a result of tragedy, can also be potent in the context of disaster relief and prevention. The notion of moderate altruism is stretched to its limit to accept and support such groups, and such acceptance should be based on self-interested rationality, i.e. as a necessity, at the very minimum. Families Against Terrorism and Extremism (FATE) talks of resilience as a way out of the pain, shame and guilt of a loved one being involved in terrorist activities by helping prevent others from getting involved. The Network of Associations for Victims of Terrorism (NAVT) has as one of its aims to promote solidarity with victims of terrorism and better citizenry. The Common Bond Project promotes solidarity among young people who are victims of terrorism.

9.3 Reasonable Self, Ethical Individualism

Alan Gewirth, a Chicago philosopher, and a neo-Kantian theorist, has elaborated the PGC (the Principle of Generic Consistency) theory which supports the self-chosen purpose of an individual, and in order for it to be achieved, states that that the individual, who is generalized as an 'agent' has what is called generic needs. The rights to generic needs are known as generic rights. It would not be rationally possible to deny these rights for oneself as an agent or for any other who displays the characteristics of an agent with a self-chosen purpose. Freedom and well-being are seen as generic needs in order for the agent to be able to act at all. Gewirth uses a dialectically necessary method to arrive at generic needs. The dialectically necessary method is as follows: If A is an agent, i.e. wanting to act for a voluntarily chosen purpose, then A should also accept (to herself, that she values the purpose so as to aspire to achieve it. By combining the dialectic method with logic, Gewirth then arrives at the position that what is true for oneself as an agent, i.e. the need to have generic needs fulfilled, should also be true for all agents. Hence a logical link is created with other agents. Gewirth also uses a dialectically contingent method to arrive at generic rights. In it, he states that the framework of human rights is predicated on the notion of generic rights. This will be picked up in the next section when discussing the state and society as a supportive state and a caring society.

So, to sum up, the PGC states that

> A person is generically inconsistent or inconsistent in a fundamental way, if, while exercising his own capacities of agency, he rejects, either in thought or in action, the possession or exercise of those capacities on the part of other persons.

Based on the PGC theory, Gewirth elaborates various concepts such as the *Reasonable Self, Ethical Individualism, Supportive State* and *Caring Society.* these concepts have been referred to earlier, and will be explained and discussed in this section. Solidarity expressed as altruistic behaviour was the subject of the discussion in the earlier section illustrating responses to disasters. This section will involve analysing solidarity as ethical individualism and altruistic behaviour as rational ethical conduct. Solidarity is further analysed in terms of its acceptable extent, under the notion of moderate altruism.

A Reasonable Self can be understood as a generically consistent agent. The notion of a Reasonable Self is a perspectival construction. It is constructed from the point of view of a community. So from the perspective of the community, the generically consistent agent in the PGC statement is a Reasonable Self. In order to understand the Reasonable Self, a comparison with the Hobbesian notion of self will help. A Reasonable Self is very different to a Hobbesian Self in that the Hobbesian Self is more atomistic in her individualism. A Hobbesian Self would be generically inconsistent and would not be exercising ethical individualism. The Reasonable Self recognises that there is a rational basis for treating others' needs as having some bearing on the fulfilment of one's own needs.

Because of the contributions made to an individual by society, the individual, to an extent, is a 'social product', which is Gewirth's *social contribution thesis* (Gewirth 1996, 83). An individual then recognizes that he or she has an obligation towards society because of society's contribution to their agency. Ethical Individualism is therefore egoistic in that 'all persons are helped to develop their abilities of agency, is regarded by each individual simply as the price she must pay to fulfil these personal rights for herself', and the 'society should be so structured institutionally that it promoted equal rights and mutuality' (Gewirth 1996, 85). Collectively, the notions of the Reasonable Self, Social Contribution thesis and Ethical individualism form Gewirthian solidarity ethics (GSE). The current framework of human rights acts as a barrier to GSE and to achieving solidarity as a policy goal. While the philosophical understanding that human rights are indivisible are embedded in the Universal Declaration of Human Rights, the European Convention has a much restricted view of rights, mainly confined to the first generation of human rights.

Therefore civil, political, social, economic and cultural rights are one but different in name. They are interrelated and indivisible. In GSE there is no need to make a reference to these differences. But from the political and legal point of view, this is essential for an understanding of how GSE can have a true harmonising impact to bring about a broad and year-round European disaster ethics policy. The artificial division between positive (social, economic and cultural) and negative rights (civil and political) is a barrier in fully realising solidarity as a policy goal. GSE too prioritises rights, but not in a similar fashion to the human rights regime. Gewirth lays out a hierarchy of needs (basic, non-subtractive and additive goods), these categories of needs do not coincide fully with the human rights notion of first, second and third generation human rights.

9.4 Mutuality as the Nature of Positive Rights

To fully understand how the Reasonable Self behaves, one must turn to the notion of mutuality. Mutuality has a definite basis in non-reciprocity in Gewirthian ethics.

While reciprocity can be understood as tit for tat – equal amounts of something exchanged between two persons – mutuality should be understood as giving or receiving something so as to secure to the other his or her generic rights. Full stop.

The other aspects that distinguish mutuality from reciprocity are:

1. In a mutual relationship, giving or receiving can be one directional, but in a reciprocal relationship, B gives something to A only if A has something to give or has given something to B.
2. In a relation of reciprocity, the beneficiaries are limited to one's prior benefactors, unlike in mutuality, where they are owed to all agents.

Because of mutuality, we see the conception that generic rights are connected in action, or in other words mutuality animates the community of rights, as an essential basis for the existence, development and flourishing of a community of agents. Mutuality puts solidarity on a rational basis. Solidarity cements a community in a lasting fashion.

Picking up on the contingent argument for human rights developed by Gewirth and also by Beyleveld and Brownsword, the argument goes that as the global community (States, individuals and other non-natural persons) has acknowledged the existence of human rights, most notably through the 1948 Universal Declaration of Human Rights and subsequently through numerous human rights treaties, it can be shown, using a dialectically contingent argument that human rights are nothing but a crystallisation of generic needs of agency (GNA). Human rights then acquire a rational foundation. Beyleveld and Brownsword (2001) argue that it follows from the acceptance of human rights that legal systems that recognise human rights must treat the PGC as a necessary criterion of legal validity, on pain of denying that they recognise human rights.

In the light of PGC, positive rights need to be fulfilled just as much as negative rights in order to create conditions for day-to-day altruism and solidarity. States risk being incongruent if they do not assume this position. Recognising positive rights is conducive to mutuality, as it supports mutuality-based solidarity behaviour by agents on a day-to-day basis. Such mundane solidarity, and more of it, is key to solidarity during disasters.

Positive and negative rights are used to distinguish and justify different types of human rights obligations owed by states to the recipients, and increasingly other entities that may have obligations under human rights law. The argument is that human rights that are negative in nature impose less onerous obligations on the State, for example for the fulfilment of the freedom of speech, it only requires that the State does not intervene. In contrast, for the fulfillment of the right to food, there is a need to invest resources on a large scale. While there are differences in how different human rights are fulfilled, that there is a qualitative difference between the two categories is not convincing. Even in the case of fulfillment of negative rights, the costs can be huge given that enforcement mechanisms need to be put in place, such as the provision for a judiciary and the police. Viewed from the Gewirthian perspective, a rational justification for positive rights provides more support for the resources spent, their core being embedded in mutuality, otherwise there is an inherent contradiction in the system. Acceptance of positive rights provides a justification for mundane solidarity to be practiced and reinforced.

The recognition of positive rights in a society makes the practice of Gewirthian solidarity ethics more natural, thereby creating conditions that make solidarity-based disaster reponse a default paradigm. This is closely reflected in the 'resilience paradigm', which spans 'from comprehensive disaster management through disaster-resistant community, disaster-resilient communities, sustainable development, sustainable hazards mitigation, invulnerable development, to comprehensive vulnerability management' (McEntire et al. 2002; Manyena 2006).

In terms of the scope of altruism, writing in his last book *Self-Fulfillment*, Gewirth (1998) argues that the standard that is set is not one of a saint, but that of a self-interested individual. Such is the expression of the trait of moderate altruism.

Ethical individualism focusses on expanding the self, while moderate altruism approaches this from the perspective of balancing the interests of the Self and the other. Prospective purposive agents or, in other words, Gewirthian actors should provide for the vulnerable while taking care of one's own needs as well as 'aspiration-fulfillment'. A fuller conception of 'aspiration-fulfillment' and its place in purposive action of agents is precluded in this short exposition of GSE.

9.5 Supportive State and Caring Society

Gewirthian actors exhibiting solidarity, comradery and altruism during a disaster are indeed behaving as moral and rational agents and their impact depends on institutional frameworks as well as their own effort. It is the duty of the State, to shape itself around such a reasonable Self to protect and promote a Reasonable Self, ever present to reinforce this idea by the development of responsive institutions. The state and society need to provide the scaffolding by being a Supportive State and a Caring Society in order to increase the chances that all agents are fulfilled in their human rights as much of the time as possible, and this includes during disasters.

At one level, being a Supportive State means that the state supports other states by supporting itself and others during times of disasters. On the other hand, as argued in Chap. 8 of this book, it means that the state has a human rights framework that it supports. Duties and obligations of States arise from generic rights that are embedded in policy as human rights. A Supportive State has an obligation to develop Positive Rights and Negative Rights. In discussing the indirect application of the PGC, Gewirth requires the transition from a minimal state to a Supportive State. The Supportive State should give more of a role for the crowd to render help during emergencies (Drury et al. 2009) so that the institutional structure that harbours Gewirthian agents supports mutualism, and not egoism (Gewirth 1996). Gewirth (1996) points out that the distribution of political power is a decisive factor, and hence the responsibility is institutional rather than individual.

The debate on negative and positive rights, or first and second generation human rights, has been too long mired in geopolitical debates between the capitalist and the socialist countries. There is a real need for a holistic conception of human rights, as inequality increases in every society and the welfare state wanes. The recognition of positive rights will provide a stronger footing for an altruistic state. The General Assembly resolution on Promotion and protection of human rights in post-disaster and post-conflict situations in 2013 requests human rights 'mainstreaming' in disaster relief in all its members states, including in Europe (UNGA 2013). In 2015, the final report of the Human Rights Advisory Committee to the United Nations General Assembly on mainstreaming human rights into disaster responses stated:

> Rights relating to the basic necessities of life (in particular relating to food, drinking water, shelter, clothing, adequate health services and sanitation), physical security and integrity (protection of the right to life and the right to be free of assault, rape, arbitrary detention and kidnapping, and threats to these rights), civil and political protection needs (rights to

religious freedom and freedom of speech, personal documentation, political participation, access to courts, and freedom from discrimination) and other economic, social and cultural protection needs (such as access to education, to receive restitution or compensation for lost property and to work) should be protected and respected through the design and implementation of concrete initiatives and mechanisms at all levels.

It is clear that the UN lays emphasis on positive and negative human rights. In the event of a CBRN incident, the application of human rights, both positive and negative, still stay important and should be equally protected. What the Gewirthian framework provides is a means of prioritising the needs, in a way that positive rights are not always disadvantaged. Gewirth lays out a hierarchy of needs known as basic, non-subtractive and additive needs. Basic needs compass features that are needed for the very possibility of action, including life itself, providing the capability to be involved in making choices and the possession of mental equilibrium sufficient to enable one to these. Non-subtractive needs are features needed to be able to act successfully, without thereby being needed for the very possibility of acting. Additive needs are features that are needed to be able to improve one's capacities for successful action, regardless of one's purposes. These needs are conceptualised hierarchically based on the possibility of successful action, therefore in the case of conflict for fulfilment of needs, rights to non-subtractive needs outweigh rights to additive needs, and the rights to basic needs outweigh non-subtractive and additive needs. A number of sustainable development goals (SDGs) enshrine objectives that will help secure basic needs. SDG 1 and 2 require states to commit to eradicating poverty and hunger, SDG 3 commits states to working to achieve good health for its citizens. Non subtractive needs include decent work (SDG 8), education (SDG 4) gender equality (SDG 5). Additive needs include participation in partnerships for success of goals, climate action, etc. The report rightly supports a pivotal role for women in disaster situations. It is clear that women play an important role in the achievement of Sustainable development goals, by playing a disproportionate role in issues such as food security. Their ability to act in a mutually reciprocal manner warrants their special status as rational agents.

Solidarity plays a role in European security policy as described in detail in Chap. 8 of this volume. Under the Treaty on the Functioning of the European Union (TFEU), Article 222 calls upon member states to work together to assist one another in the event of a terrorist attack, a man-made disaster or a natural disaster. While this is a necessary inclusion in the policy, there is room for criticism. The reference to a 'spirit of solidarity' in Article 222 can be seen as a weak provision, with the potential to be trumped by weaker values. However the reference to joint action is refreshing and acts as a mellowing factor to a potentially watered down solidarity requirement. GSE would require that post-disaster recovery should continue to see the Supportive State in action. Konstadinides (2013) rightly argues that the positioning of Article 222 in the Treaty is indicative of its limited significance, casting doubts on solidarity as a policy goal. A clearer approach to Article 222 would be to read 'the solidarity clause in terms of a kind of mutual insurance tool able to minimize the risks at a common minimal cost but not in any case an altruistic tool created in order to provide aid for those countries that are not able to face the challenges posed by major crisis situations on their own' (Chap. 8 of this book, page 5).

It is in the remit of the Supportive State to not hinder, but support the development of caring communities. Disaster relief can be called for unexpectedly and in unknown scales, therefore having caring communities to call upon can provide for fulfillment of agents' needs. However, CBRNE incidents bring out rational and moral behaviours of individuals irrespective of a widespread caring community that exists, and this is a point made earlier through examples of actual incidents of bombings in major European capitals, terrorist shooting in Paris and chemical attacks in Iraq and Tokyo. In order to address the peculiarities of the relationship between CBRNE, mutuality and solidarity, three points need to be made. First, that the incident and the immediate aftermath attract the most mutuality underpinned actions. Second, the knowledge that it is a CBRNE incident is essential to kindle the mutualistic propensity. Finally, solidarity in its deepest form, i.e. mutuality has more chance of being spontaneously displayed during a CBRNE event than during other events that triggers altruistic behaviour.

Caring communities are motivated and organized (Cretney 2016). It is interesting that in some less ordered societies, state officials (acting as if in their private capacity) can exhibit much more rational ethical behaviour than in the case of officials of developed States. This is perhaps based on cultural values that pervade strongly in many societies. Cultural values in societies can play a strong role in ethical behaviour for resilience (Kenney and Phibbs 2015). The role of social organisations in organizing relief has become increasingly better and important in the context of disaster relief. Groups have shown independence in defining the resilience framework they need. In this context, communities need to proactively work to infuse 'caring' and shape what this means for their work. There is lack of clarity in the indicators for community resilience because of the neglected status of community resilience (Manyena 2006). Central to Gewirth's conception of caring communities is his *social contribution thesis*. This thesis should help apportion obligations and shape appropriate indicators for community resilience. Community resilience should work better on mutualistic foundations as was demonstrated in the Dunedin earthquake mobilization in a locality that was badly hit (Manyena 2006). Sustainable Development Goal 11 provides a mutualistic framework, in keeping with Gewirthian social contribution thesis. By 2030, it aims to enhance capacity for participatory human settlement planning and management in all countries, thereby providing individuals the opportunity to give back to the communities and fulfil their role as mutual agents (SDG 2015). All members of society are enrolled into social contributions, except that the social contribution thesis does not operate in its full scope for those born into extreme deprivation, because unless they are helped by society first, they are not able to become productive agents, to then take on responsibilities.

The right to development (RtD) is pertinent in the discussion of caring communities. RtD is a controversial third generation human right. Gewirth stakes a claim via the social contribution thesis for a 'collective right of the community to an institutionalised system of support from those it has benefited', which could be very much likened to RtD. This is an interesting way to demonstrate a caring community, i.e. by demanding something. But that is the reality and complexity of communities. The caring community gives as well as takes, so as to be able to give, and take!

Caring communities play a role in reducing the impact of future disasters. The adaptive capacity of the caring communities is greater than that of non-caring communities. Pre-disaster community activities contribute to better community resilience, and these are examples of Gewirthian caring communities in action, operating with heightened awareness that disasters are part of modern societies. Social support and social participation have been identified as elements of pre-disaster societies that have helped respond and adapt post-disaster (Cretney 2016). Working towards SDG 11 to make cities inclusive, safe, resilient and sustainable would go a long way in realising Gewirthian values. A key target, by 2020, is to substantially increase the number of cities and human settlements adopting and implementing integrated policies and plans towards inclusion, resource efficiency, mitigation and adaptation to climate change, resilience to disasters, and develop and implement, in line with the Sendai Framework for Disaster Risk Reduction 2015–2030, holistic disaster risk management at all levels (SDG 2015).

9.6 Conclusion

Construing solidarity as Gewirthian mutuality in disaster ethics creates a rational human rights framework that operates as a ‘community of rights’. It provides a worthy framework during a difficult situation where moral decisions are taken within a structure already made conducive to human rights and human dignity. Mutuality is a key human rights concept, manifesting in the recognition and promotion of positive human rights law in a Supportive State and provides continuity between resilience and disaster paradigms. The role of communities in disaster relief and resilience is often ignored or only grudgingly tolerated, however Gewirthian solidarity ethics places the community in an integral position for the fulfilment of human rights. This chapter seeks to reconcile altruistic behaviour with a human rights framework that recognises and builds upon this phenomenon.

References

Barber, Rebecca J. 2008. Protecting the right to housing in the aftermath of natural disaster: Standards in International Human Rights Law. *International Journal of Refugee Law* 20 (3): 432–468. https://doi.org/10.1093/ijrl/een024.

Bean, Hamilton. 2011. No more secrets: open source information and the reshaping of U.S. Intelligence. Praeger Security International.

Beyleveld, Deryck, and Roger Brownsword. 2001. *Human dignity in bioethics and biolaw*. Oxford: Oxford University Press.

Council of the European Union. 2009. Council conclusions on strengthening chemical, biological, radiological and nuclear (CBRN) security in the European Union – An EU CBRN Action Plan.

http://register.consilium.europa.eu/doc/srv?l=EN&f=ST%2015505%202009%20REV%201. Accessed 11 Apr 2016.
Cretney, R.M. 2016. Local responses to disaster. *Disaster Prevention and Management* 25 (1): 27–40.
Donald, I., and D. Canter. 1992. Intentionality and fatality during the Kings Cross underground fire. *European Journal of Social Psychology* 22: 203–218.
Drury, John, Chris Cocking, and Steve Reicher. 2009. The nature of collective resilience: Survivor reactions to the 2005 London bombings. *International Journal of Mass Emergencies and Disasters* 27 (1): 66–95.
Fischer, Henry W., III. 2000. Mitigation and response planning in a bio-terrorist attack. *Disaster Prevention and Management: An International Journal* 9 (5): 360–368.
Gewirth, Alan. 1978. *Reason and morality*. Chicago: The University of Chicago Press.
———. 1996. *The community of rights*. Chicago: The University of Chicago Press.
———. 1998. *Self-fulfillment*. Princeton: Princeton University Press.
Griswold, Alisha. 2013. Digital detectives and virtual volunteers: Integrating emergent online communities into disaster response operations. *Journal of Business Continuity & Emergency Planning* 7 (1): 13–25.
Hart, J., and P. Clevestig. 2009. Reducing security threats from chemical and biological materials. In *SIPRI yearbook: Armaments, disarmament and international security*, 413–433.
Kenney, M.C., and S. Phibbs. 2015. A Māori love story: Community-led disaster management in response to the Ōtautahi (Christchurch) earthquakes as a framework for action. *International Journal of Disaster Risk Reduction* 14: 46–55.
Konstadinides, Theodore. 2013. Civil protection cooperation in EU law: Is there room for solidarity to wriggle past? *European Law Journal* 19 (2): 267–282.
Manyena, Siambabala Bernard. 2006. The concept of resilience revisited. *Disasters* 30 (4): 433–450.
McEntire, D.A., Christopher Fuller, Chad W. Johnston, and Richard Weber. 2002. A comparison of disaster paradigms: The search for a holistic policy guide. *Public Administration Review* 62 (3): 267–281.
Murakami, Haruki. 2000. *Underground: The Tokyo gas attack and the Japanese psyche*. London: Harvill Press.
Pangi, Robyn. 2002. *Consequence management in the 1995 Sarin attacks on the Japanese subway system*. http://belfercenter.ksg.harvard.edu/files/consequence_management_in_the_1995_sarin_attacks_on_the_japanese_subway_system.pdf. Accessed 11 Feb 2016.
Pfaff, Donald W. 2015. *The altruistic brain: How we are naturally good*. Oxford: Oxford University Press.
United Nations General Assembly. 2013. Promotion and protection of human rights in post-disaster and post-conflict situations, A/HRC/RES/22/16.
———. 2015. *Final research-based report of the Human Rights Council Advisory Committee on best practices and main challenges in the promotion and protection of human rights in post-disaster and post-conflict situations*. http://www.ohchr.org/EN/HRBodies/HRC/AdvisoryCommittee/Pages/HRpostdisasterandpostconflictsituations.aspx. Accessed 14 Apr 2016.
United Nations, Sustainable Development Goals. 2015. http://www.un.org/sustainabledevelopment/cities/. Accessed 12 Nov 2017.

Chapter 10
Clinical Trials in Crisis Situations: Ethical Issues

Miguel A. Ramiro Avilés

> *Desperate times do not require desperate morality just different morality (...) Desperation is a reason to think more, not less, about ethics*
>
> Arthur Caplan

Abstract The implementation of a clinical trial within a crisis situation – including chemical, biological, radiological, nuclear and explosive (CBRNE) events triggered by natural causes or bioterrorist attacks – raises issues related to the balance between public health and human rights. In such a scenario we need to ponder ethical principles and legal rules, especially when the agents causing illness or death in people, animals or plants do not respond to current standard of care or no approved treatments exist. Public health crises complicate ethical concerns common to clinical research and pose challenges to the way ethical principles for human subjects research are interpreted and applied. It is necessary to give careful thought to which normative requirements (both legal and ethical) should be satisfied in order to carry out a clinical trial in a public health crisis. Double normative standards and mandatory participation are issues of particular importance to the design and implementation of clinical trials in public health crises.

Keywords Public health crisis · Clinical trials · Research ethics · Rights · Double standard · Mandatory participation

M. A. Ramiro Avilés (✉)
University of Alcalá, Madrid, Spain
e-mail: miguelangel.ramiro@uah.es

D. P. O'Mathúna, I. de Miguel Beriain (eds.), *Ethics and Law for Chemical, Biological, Radiological, Nuclear & Explosive Crises*, The International Library of Ethics, Law and Technology 20, https://doi.org/10.1007/978-3-030-11977-5_10

10.1 Introduction

The implementation of a clinical trial within a crisis situation raises issues related to the balance between public health and human rights and 'the question of whether clinical research in acute epidemic situations ought to occur at all, and, if so, under what ethical or moral conditions and constraints' (Ezeome and Simon 2010, 2). It is a scenario with two overlapping dynamics, one inherent in the public health crisis and the other inherent in the research enterprise (London 2009).

A public health crisis is a situation where a disease spreads quickly and extensively causing a real, serious and imminent risk to public health by affecting the health and physical integrity not only of the members of a particular community but also globally. This exceptional situation means that governments, respecting the rule of law, need to take certain actions that limit, restrict or impinge on the exercise of citizens' rights and freedoms (London 2009). In crisis situations where public health is affected, vaccination or therapeutic treatments may be made mandatory; freedom of movement may be limited through quarantine or forced institutionalization; and contact maps may be drawn up or a cordon sanitaire established. These are all classic public health measures and little debate occurs over their necessity in certain circumstances (Hodge et al. 2014); 'they are not high-tech, but they have a proven track record of controlling infectious outbreaks' (Rid and Emanuel 2014, 1896).

If the viruses, bacteria or agents have a proven treatment (the standard of care), we may suppose that therapeutic treatments, vaccination or some other public health measures may be imposed. Governments commonly pass public health laws that include such exceptional measures (mandatory vaccination or treatment) for exceptional situations (public health crises), what is called the *democratic reason of state* (Thompson 1987). But sometimes viruses, bacteria and other agents are resistant to current medicines or no approved treatments exist. Such a scenario brings the second dynamic to the fore because it will be necessary to design a clinical trial in order to discover or test the best therapeutic treatment and at the same time to protect the rights, safety and well-being of the participants. As Alex John London states, 'inherent in the clinical trial enterprise is the potential for conflict between the interests of current participants and the interests of the future beneficiaries of that research (…) they bear the risks and burdens associated with purely research-related procedures as part of an effort to generate information that will ensure that future patients receive a better standard of care' (London 2009, 1175–1176).

Public health crises complicate ethical concerns common to clinical research and pose challenges to the way ethical principles for human subject research are interpreted and applied (Presidential Commission 2015). Moreover, most countries lack regulatory frameworks permitting the use of unapproved interventions on humans in crisis scenarios. For instance, article 35 of the European Union Regulation on Clinical Trials only regulates clinical trials in emergency situations. Recently, the Council for International Organizations of Medical Sciences (CIOMS) revised its *International Ethical Guidelines for Health-related Research Involving Humans* and added Guideline 20 on research in disasters and disease outbreaks. This new guideline states that 'health-related research should form an integral part of the disaster response' and at the same time 'the conduct of research must not unduly impact the response to the victims of a disaster' (CIOMS 2016, 75). Conducting research during disaster and disease outbreaks raises important challenges, which 'need to be carefully balanced with the need to ensure the scientific validity of the research and uphold ethical principles in its conduct' (CIOMS 2016, 75).

Careful thought must be given to which normative requirements (both ethical and legal) should be satisfied in order to carry out a clinical trial in a public health crisis, bearing in mind that different infectious diseases require different forms of response. Scientific differences have implications for 'how the disease can and should be addressed through clinical medicine, public health surveillance and intervention, legal provisions, and so forth. (…) Legally, there may be significant differences in the obligations to intervene depending on the infectious disease in question' (Smith and Silva 2015, 136).

An Ebola outbreak is not merely a biological problem but also a moral challenge because foundational moral considerations should play a key role to bring Ebola under control in an expeditious and fair manner (Benatar 2015; Rothstein 2015), and it could be used as a model because there are now ten or so vaccines and treatments at various stages of research, development and clinical testing (Strauss 2014). Moreover, the World Health Organization expert panel recommendation stated that in the context of a public health crisis 'it would be acceptable on both ethical and evidential grounds to use as potential treatments or for prevention unregistered interventions that have shown promising results in the laboratory and in animal models but have not yet been evaluated for safety and efficacy in humans, provided that certain conditions are met' (WHO 2014, 1). This recommendation fuelled the debate over whether it is acceptable to offer interventions on limited evidence (Shah et al. 2015) and raised the question of which ethical standards, if any, ought to govern the testing of unregistered interventions (Caplan et al. 2015). In short, the recommendation broached the issue of whether some regulatory flexibility is reasonable (Goodman 2014) because 'in a humanitarian emergency – as in situations in which people face imminent death or disability and seek access to unapproved agents – flexibility in ethical standards is required' (Caplan 2015, e16).

Clinical trials in crisis situations, like the 2014–2016 Ebola outbreak, must be carried out if there is no proven therapeutic treatment or vaccine, but at the same time the rights and wellbeing of all participants must be protected. We need to strike a balance between these two overlapping ideas, while the final outcome must ensure 'that the burdens of the research are not disproportionately borne by persons who are already socially, economically, or politically disadvantaged and that special protections are in place to ensure that the risks to individuals from such groups are minimized' (London 2009, 1201).

This paper has two interconnected parts, and concludes with some brief final ideas. The first part examines the principles that are fundamental for a clinical trial's implementation to be ethically acceptable. The second part explores two specific problems attached to the design of a clinical trial in a public health crisis situation: normative double standards and mandatory participation. The approach taken in this chapter complements that in Chap. 11 where solidarity is used as an underlying ethical principle and value to address the ethical challenges of research in CBRNE crises.

10.2 How Can a Clinical Trial Be Carried Out Ethically?

According to article 2 of the EU Regulation on Clinical Trials, a clinical trial is 'any investigation in relation to humans intended: (a) to discover or verify the clinical, pharmacological or other pharmacodynamic effects of one or more medicinal products; (b) to identify any adverse reactions to one or more medicinal products; or (c) to study the absorption, distribution, metabolism and excretion of one or more medicinal products; with the objective of ascertaining the safety and/or efficacy of those medicinal products'. All clinical trials must also fulfil 'any of the following conditions: (a) the assignment of the subject to a particular therapeutic strategy is decided in advance and does not fall within normal clinical practice; (b) the decision to prescribe the investigational medicinal products is taken together with the decision to include the subject in the clinical study; or (c) diagnostic or monitoring procedures in addition to normal clinical practice are applied to the subjects' (European Union 2014). In short, a clinical trial of medicinal products for human use is an invasive procedure carried out for the purposes of research which may jeopardize the health, life and physical integrity of the participants. It is an exceptional procedure because it makes use of experimental agents in the absence of human safety, efficacy, or dosing data, in the context of a scientific protocol with the goal of benefiting not an individual patient (as with compassionate use) but society at large and future patients (Miller and Brody 2003). 'It is widely accepted that randomization is ethical when clinical equipoise is established, that is, when there is genuine uncertainty whether an experimental treatment has benefits or risks that exceed those of conventional care. Equipoise, however, breaks down when available conventional care offers little benefit, some agents appear promising and safe, and mortality is extremely high' (Caplan et al. 2015, 6).

This chapter will not address the place for compassionate use or expanded use in CBRNE events or public health crises. Compassionate use 'refers to the use of an unapproved agent, outside the context of a scientific protocol, with the goal of benefiting an individual patient with a serious, usually life-threatening condition' (Joffe 2014, 1299). Confronted with a disease like Ebola, the inclination towards compassionate use is understandable because we want to save people's lives, as a manifestation of the rule of rescue (Edwards 2013). 'Where a remotely possible treatment is available for a likely fatal disease, no matter how risky, the sick will almost always opt for it. That is why there has been a push (...) for compassionate use and expanded access to experimental drugs, devices, and biologics for the terminally ill' (Caplan et al. 2015, 6). However, 'a compassionate use approach will not necessarily prevent more deaths than would administration of the drug in a well-designed clinical trial' (Joffe 2014, 1299). In this connection, some commentators urge policy makers to organize appropriate clinical trials (Joffe 2014; Cox et al. 2014; Rid and Emanuel 2014) and relax the regulatory restrictions (Edwards 2013).

A widely accepted ethical framework proposes eight requirements that systematically elucidate criteria for evaluating the ethics of clinical research studies (Emanuel et al. 2000, 2004). This framework ensures 'that research can generate the information that will enhance the capacity of the health infrastructure of a community to respond to a particular threat, but it should also ensure that research is carried out in a way that embodies, and communicates to the public, certain facts about the basic social and governmental institutions of the community' (London 2009, 1174). The requirements 'are meant to guide the ethical development, implementation, and review of individual clinical protocols (…) [the requirements] are intended to elucidate the ethical standards specific for clinical research and assume general ethical obligations' (Emanuel et al. 2000, 2702). They can also help to anticipate potential and important ethical issues in decision making (see Chap. 11). The ethical requirements are not limited to 'normal' situations because they can also be applied to exceptional situation as public health crisis but their application requires adaptation to because 'the same basic principles of morality appropriate in all medical cases and research also apply, but some morally problematic actions may be excused, or even justified, because the circumstances are not usual; and the ways in which they are unusual may be morally relevant' (Ezeome and Simon 2010, 2).

1. *Collaborative partnership*. In clinical trials a collaborative partnership between researchers, sponsors, policy makers, representatives of patients and community leaders helps to minimize the possibility of exploitation by determination of 'whether the research is acceptable and responsive to the community's health problems' (Emanuel et al. 2004, 932). This partnership also develops the capacity to respect community values, culture, traditions and social practices (Ezeome and Simon 2010). But this collaborative partnership needs time to be developed, so 'acute epidemic research therefore needs forethought and anticipatory planning before the onset of the epidemic, since once it sets in there will be no time for all these' (Ezeome and Simon 2010, 5). In this sense, 'time between major disasters is ideal for deliberative thinking that makes for good planning, for laying the groundwork for future efforts, and ultimately for good results of scientific inquiry' (Lurie et al. 2013, 1253). As CIOMS Guideline 20 states, 'research in disaster and disease outbreaks should ideally be planned ahead. Health officials and research ethics committees should develop procedures to ensure appropriate, expedient and flexible mechanisms and procedures for ethical review and oversight' (CIOMS 2016, 76).
2. *Social value*. The research must provide social value by direct or indirect enhancement of knowledge about a particular disease or health condition. 'Only if society will gain knowledge, which requires sharing results, whether positive or negative, can exposing human subjects to risk in clinical research be justified' (Emanuel et al. 2000, 2703). According to the Commentary on CIOMS Guideline 1, 'social value refers to the importance of the information that a study is likely to produce. Information can be important because of its direct relevance for

understanding or intervening on a significant health problem or because of its expected contribution to research likely to promote individual or public health' (CIOMS 2016, 1). Therefore, participation of human beings in clinical trials is unacceptable if no worthwhile information is to be generated. In crisis situations like the Ebola outbreak or the HIV epidemic in the 1990s, where no treatments had been shown to be safe and effective but there was an urgent need to identify therapies, clinical trials can be a useful means of obtaining them. The lack of scientific evidence 'provides ethical justification for research, even randomly assigning research subjects to different interventions' (O'Mathúna 2015, 32).

3. *Scientific Validity*. The design of the clinical trials must be methodologically rigorous and the most appropriate for the research question. The *null hypothesis* requires 'controversy within the scientific community about whether the new intervention is better than standard therapy, including placebo, either because most clinicians and researchers are uncertain about whether the new treatment is better, or because some believe the standard therapy is better while others believe the investigational intervention is superior' (Emanuel et al. 2000, 2704). A *null hypothesis* contributes to the protection of participants not exposing them to obsolete interventions with no scientific doubt. The design of the clinical trial should not cause harm to research subjects by exposing them to unnecessary risks in participation (O'Mathúna 2015). As the CIOMS Guideline 1 states, 'proposed studies [must be] scientifically sound, build on an adequate prior knowledge base, and are likely to generate valuable information' (CIOMS 2016, 1).

A randomized clinical trial (RCT) with one control group (standard of care, including placebo) is considered the scientific gold standard but may not be the best design in a public health crisis. During the 2014–2016 Ebola outbreak two different stances were adopted on RCTs. One view held that the priority for clinical research should have been to identify safe and effective interventions as efficiently and reliably as possible, and that RCTs were the best way to achieve this (Joffe 2014; Rid and Emanuel 2014; Rid 2015; Shaw 2014). The other view held that priority should be given to provide access to the potential benefits of experimental interventions to as many participants as possible using valid research designs (Caplan et al. 2015; Adebamowo et al. 2014; Cooper et al. 2015). Although it may seem to be a merely technical matter of scientific methodology, the choice of one strategy or another may have profound implications for the participants' well-being and rights.

4. *Fair Subject Selection*. Scientific objectives, not vulnerability or privilege, and a favourable risk-benefit ratio should determine the inclusion criteria for the communities as study sites and for individuals as research subjects (Emanuel et al. 2000, 2704). As CIOMS Guideline 3 states, 'groups, communities and individuals invited to participate in research must be selected for scientific reasons and not because they are easy to recruit because of their compromised social or economic position or their ease to manipulation' (CIOMS 2016, 7). Transparency should exist in selection criteria that are consistently applied to avoid categorical exclusions that are discriminatory. In this sense, groups or individuals 'should not be excluded from the opportunity to participate in research without a good

scientific reason or susceptibility to risk that justifies their exclusion' (Emanuel et al. 2000, 2704).

The fair selection of participating subjects implies: the prohibition of direct, non-direct and by association discrimination against anyone; compliance with principles of equality of treatment and of opportunities; only those who satisfy the inclusion and exclusion requirements as stipulated in the research protocol may become subjects of the research; and only those who are strictly necessary in order for the research question posed in the research protocol to be answered must be enrolled as participants. In this sense, CIOMS Guideline 15 requires that 'when vulnerable individuals and groups are considered for recruitment in research, researchers and research ethics committees must ensure that specific protections are in place to safeguard the rights and welfare of these individuals and groups in the conduct of research' and protections can include 'that the research be carried out only when it is targeted at conditions that affect these groups' (CIOMS 2016, 57).

In the Ebola outbreak there was controversy over how to select a study population due to the lack of experimental drugs. The allocation of scarce resources meant that the selection of subjects became an ethically important and complex issue. One argument was that 'we owe a duty to those workers who have knowingly placed themselves at risk of disease and death in order to serve the needs of others (…) Moreover, there is a practical reason to consider treating healthcare workers first. It serves the interests of the majority of patients to keep medical workers and caretakers in the field' (Donovan 2014, 3). But the opposite argument was that 'supportive care should be directed at those with few comorbid conditions and early disease manifestations due to higher rates of recovery and lower rates of transmission to personnel' (Hantel and Olusola Olopade 2015, 142).

5. *Favourable Risk-Benefit Ratio*. Within the context of standard clinical practice and the particular research protocol, risks must be minimized, potential benefits enhanced, and the potential benefits to individuals and the knowledge gained for society as a whole must outweigh the risks (Emanuel et al. 2000). According to CIOMS Guideline 4, 'before inviting potential participants to join a study, the researcher, sponsor and the research ethics committee must ensure that risks to participants are minimized and appropriately balanced in relation to prospect of potential individual benefit and the social and scientific value of the research' (CIOMS 2016, 9). The evaluation of the *ratio* is a process in which risks and potential benefits must be evaluated both individually and as an aggregate in order to minimize the risks and boost the potential benefits. This evaluation is one of the most complex requirements to achieve (Rid and Wendler 2011). Its relevance is partly due to the interconnection with the principle of autonomy in order to assess the level of risk one person can take on. The 'minimal risk' standard is one of the most used but 'when thousands of people are confronted with a life-threatening disease, and no specific therapies or preventive measures exist, it can be ethically acceptable to assume greater risks' (Rid and Emanuel 2014, 1896). Martin Delaney graphically explained this for the HIV case: 'It is as if I am in a disabled airplane, speeding downward out of control. I see a parachute hanging

on the cabin wall, one small moment of hope. I try to strap it on, when a government employee reaches out and tears it off my back, admonishing, "You cannot use that! It does not have a Federal Aviation Administration sticker on it. We do not know if it will work"' (Delaney 1989, 416). In the Ebola outbreak, experimental interventions might have had various toxic risks, yet to persons with Ebola, these may have seemed irrelevant or even absurd in comparison with the natural history of the disease and the local standard of care (Caplan et al. 2015)

6. *Independent Review*: Research ethic committees have become key players in the development and governance of clinical trials because they provide an external and independent source of oversight designed to verify that the pursuit of the goals of the clinical trial do not overlook the protection of certain basic values and principles related to human rights. According to CIOMS Guidelines 8 and 23, all health-related research involving human participants must be 'reviewed ethically and scientifically by competent and independent research ethics committees' which is 'critical to engender community trust for research' (CIOMS 2016, 29) and to protect the rights, safety and well-being of participants. Research ethics committees provide review by the qualified opinions of its members (healthcare professionals, representatives of patients' associations, lawyers, bioethicist, representatives of relevant advocacy groups and laypersons) and guide the conduct of private individuals, the decisions of healthcare workers and researchers, thereby minimizing the potential impact of conflicting interests (Morin et al. 2002). Research ethics committee reviews are indispensable for socially responsible research, acting as watchdogs to supervise that the people enrolling in clinical trials are properly informed, not subjected to unacceptable risks, properly compensated in case of research-related injuries and not exploited.
7. *Informed Consent*. Voluntary consent is absolutely essential, which means there is an *inalienable right* to refuse or withdraw (McConnell 2010). Potential participants should be informed in culturally and linguistically appropriate formats about the research and provide their voluntary consent. 'The purpose of informed consent is two-fold: to ensure that individuals control whether or not they enrol in clinical research and participate only when the research is consistent with their values, interests and preferences' (Emanuel et al. 2000, 2705). With the aid of informed consent, research participants are treated as moral agents worthy of equal concern and respect, and are not reduced to 'mere means' for the benefit of others. Clinical research requires that participants receive more information than for medical care, including information about the subordination of therapeutic interests to scientific interests. Research subjects should be aware that clinical researchers may make decisions that do not prioritize their best therapeutic interest but rather the success of the research and the benefit of future patients in order to avoid therapeutic misconception (Appelbaum et al. 1982). The cultural adaptation means that in some contexts community and family consent procedures must be implemented (Diallo et al. 2005), but the community 'will need special education on the fact that the consent is primarily for individual patients to give' (Ezeome and Simon 2010, 9). As CIOMS Guideline 9, information must be reviewed 'if there is a substantial change in the conditions or procedures of

the research, or if new information becomes available that could affect the willingness of participants to continue' (CIOMS 2016, 33). Informed consent then should be seen as an ongoing process.

In some cases, certain potential research subjects are not capable of giving informed consent. In these situations, if reasonable accommodations of the informed consent process have proved futile, according to CIOMS Guideline 10, legally authorised representatives can give informed consent (CIOMS 2016, 37).

Information and voluntary consent pose a challenge in disaster settings (O'Mathúna 2015) mainly because potential participants may be considered vulnerable, either because obstacles exist to obtaining informed consent ('consent-based vulnerabilities'), because participation may increase the participants' exposure to risk ('risk-based vulnerabilities'), or because there is an unequal distribution of benefits and burdens ('justice-based vulnerabilities) (Coleman 2009). In this sense, 'obtaining valid informed consent is the most encumbered area in acute epidemic research [because] the overwhelming desire to get treatment may compel individuals to enter research studies without due consideration of the nature of the study, is risks and benefits, and other factors' (Ezeome and Simon 2010, 9).

8. *Respect for Potential and Enrolled Subjects*: All persons involved in clinical trials must be protected and respected, which entails different activities: protection of privacy, confidentiality of personal data and recognition of the habeas data rights; refusal and withdrawal without penalty; receiving new information arising in the course of the research in order to update consent; monitoring of data and human well-being throughout participation and appropriate treatment if subjects experience adverse reactions; information about what was learned from the research and post-trial access to medicines prior to commercialization; and, last but not least, fair compensation for research-related injuries under a non-fault system (Ramiro Avilés 2015).

10.3 Double Standards and Mandatory Participation in Clinical Trials in Crisis Situations and CBRNE Events

The above ethical principles for clinical trials will now be applied to two issues of particular ethical importance in the design and implementation of clinical trials in public health crises and CBRNE events. The first is the normative double standard, which may arise when the clinical trial is carried out globally and therefore affects countries with asymmetrical levels of protection in terms of the right to health. The second is the normative flexibility required by situations such as a public health crisis where mandatory participation in a clinical trial might even become another public health measure.

10.3.1 The Normative Double Standard in Globalized Clinical Trials in Crisis Situations

When a clinical trial will be conducted in developed and developing countries, an ethical challenge may arise. In some situations, participants in developing countries may have a lower level of protection than participants in developed countries because of differences in how ethical and legal protections are applied and enforced. Such variability in the application of clinical research ethical principles has been called the "normative double standard". Some variability (ethical pluralism) can be accepted due to the different cultural contexts. However, such variability should not be accepted where it would be ethically unacceptable to conduct a specific research protocol in the developed country (Macklin 1999). We see acceptable variation where in some countries community leaders are engaged in the information consent process. Community participation is based on respect for community autonomy, people's right to self-determination, due consideration of the consequences of the research for the community as a whole, and a more complete understanding of human autonomy (Buchanan et al. 2008). However, participation of communities in the informed consent process should be accepted only if women's opinions and decisions are not excluded because, according to Commentary on CIOMS Guideline 7, 'when community leaders are men only, researchers should actively include the views of women' (CIOMS 2016, 25).

This classical ethical dilemma over double standards has been debated by scholars for decades (Schüklenk 2014). According to Ruth Macklin, different standards should not be accepted (Macklin 2004). Nevertheless, some do accept double standards. Macklin quotes one researcher from the Uganda Cancer Institute defending different standards when 'appropriate authorities, including the national ethics review committee, have satisfied themselves that the research meets their own ethical requirements' (Macklin 2004, 3). From the perspective of *realpolitik*, in its report *Ethics and Ebola,* the Presidential Commission for the Study of Bioethical Issues has also come to the same position.

During the 2014–2016 Ebola crisis, two distinct positions on the design of clinical trials emerged. Proponents of (RCTs) designed with a control group (standard of care, including placebo) argued that subjects in the control group should receive the best available supportive care (Rid 2015). This begs the question of how to interpret *best available*: 'On the one hand, it might be interpreted as meaning the *best possible* supportive care, which would require the more sophisticated interventions that can be provided in well-resourced intensive care units (such as renal dialysis and ventilators) that are rarely available in the context of the current Ebola epidemic. On the other hand, the term might be interpreted to mean local de facto care wherever the specific trial is occurring, which might be very little' (Presidential Commission 2015, 39). According to the Presidential Commission, 'providing supportive care of the level provided to the few Ebola patients treated in hospitals in the United States and Europe is both unrealistic and impractical and could not be sustained by many health systems most affected by this epidemic (…) the most appropriate comparator for an experimental Ebola treatment in this context is the best supportive care available and sustainable in the community in which the research is conducted and where the intervention will be used' (Presidential Commission 2015, 39). The first interpretation 'holds that the relevant standard of care is the one determined by the best therapeutic methods available anywhere in the world. Call this the global reference point'; the second 'holds that the relevant standard of care is determined by the standard that prevails in the country in which the trial is conducted. Call this the local reference point' (London 2000, 382).

This debate, a re-run of the post HIV pregnant women clinical trial controversy (Lurie and Wolfe 1997; Angell 1997), reveals a fundamental misunderstanding of the rights and welfare of the research participants in the Ebola outbreak and exacerbates pre-existing inequalities, most of them created by neoliberal policy changes that 'have left many developing states without the health resources and infrastructures necessary to respond to the majority of the world's disease burden' (Meier and Mori 2005–2006, 109).

We should 'demand that whatever has been made available for rescue [European and American citizens be used to try to rescue infected West Africans in format similar to what was tried in wealthy nations. To do otherwise is to violate the requirements of fairness that like cases ought to be treated alike' (Caplan et al. 2015, 5). The normative double standard ought to be challenged and it will be ethically unacceptable to design clinical trials in global public health crisis where the standard treatment with which the experimental medicinal product was to be compared is based on a criterion (participant's residence) that should not be relevant.

In situations like these, the local reference point would harm participants because the due standard of care against which the clinical trial operates cannot be the *de facto* standard of care in the developing country (Larkin 2015). In a global public health crisis, the standard of care against the clinical trial operates should be the *de jure* standard of care, which is set not by the actual practice but 'by the judgment of experts in the medical community as to which diagnostic and therapeutic practices have proven most effective against the illness in question (London 2000, 384).

It is not enough to ensure that those living in poor countries have access to whatever (sub-optimal) treatments they may be entitled to in normal circumstances; there must also be guarantees that those caught up in a global public health crisis have global access to the best available treatment anywhere in the world. Moreover, once the clinical trial is completed, and according to the post-trial access plans stated in Commentary to CIOMS Guideline 2, they should also have unrestricted and free access to the outcomes of the research, including the medicinal product or treatment developed from the research. Those who are suitable participants in a clinical trial cannot be unsuitable beneficiaries of its results (Yearby 2003–2004).

10.3.2 Mandatory Participation

Mandatory participation in a clinical trial during a public health crisis or CBRNE event is an issue that runs counter to basic ethical principles since it jeopardizes the Kantian means-ends rationale: 'to enrol individuals in clinical research without their authorization is to treat them merely as a means to purposes and ends they may not endorse and deny them the opportunity to choose what projects they will pursue' (Emanuel et al. 2000, 2706). It also runs counter to all legal rules which codify that ethical principle, for example, article 7 of the International Covenant on Civil and Political Rights: 'No one shall be subjected to torture or to cruel, inhuman or degrading treatment or punishment. In particular, no one shall be subjected without his free consent to medical or scientific experimentation'; article 6.2 of the Universal Declaration on Bioethics and Human Rights: 'Scientific research should only be carried out with the prior, free, express and informed consent of the person concerned'; or article 3.2.a of the Charter of Fundamental Rights of the European Union: 'In the fields of medicine and biology, the following must be respected (…) the free and informed consent of the person concerned, according to the procedures laid down by law'. Participation in a clinical trial must be always voluntary, and that requires coherent procedures for transmitting information and obtaining consent.

However, requirements and interpretations of research ethics principles should be more flexible in public health crises because we are facing an exceptional situation. If we are able to find a balance between participants' rights and the methodological requirements of the research, then we are in a good starting position to relax some ethical principles and legal rules, and to justify that research without consent can be done ethically (Gelinas et al. 2016).

In a public health crisis, proven and effective vaccination or therapeutic treatments may be imposed because they form part of the standard of care which aims to benefit the individual patient and society as a whole. We could ask if participation in a clinical trial may also be made mandatory, like other public health measures may be, even though a clinical trial is not intended to provide someone with the best, proven diagnostic or therapeutic treatment, but to obtain generalizable information for use in the development of vaccines or medicinal products for future patients.

An affirmative answer to this question will not be based on an absolute moral imperative, but can be based on utilitarian premises. We need to take into account the consequences of the decisions adopted: 'many people would die, and quickly' (Edwards 2013, 7). If participation in a clinical trial were not mandatory and informed consent had to be obtained, the upshot could be insufficient participants and the clinical trial would yield invalid findings.

In a public health crisis caused by a virus, bacteria or agent resistant to conventional standard of care, even to the most sophisticated interventions, mandatory participation in clinical trials should be considered among the range of public health measures that may be necessary. A measure like this would have to 'identify special needs of various specific communities and to formulate strategies for the conduct of research that is responsive to their needs (…); clarify and address the ethical questions that arise during public health emergencies (...); manage a risk to the public trust on which all human research ultimately depends – the risks that scientists will be perceived as exploiting people or communities who are in their most vulnerable state during a disaster' (Lurie et al. 2013, 1253).

However, mandatory participation must be legally justified, passed in parliament and submitted to public debate and democratic control. Legislation regarding clinical trials already permits exceptions to be made in emergency situations where due information and consent may be waived because the consent is taken for granted and is expected to be given once the circumstances that temporarily impede receipt of the information and granting of consent have disappeared (e.g. art. 35 of the European Union Regulation on Clinical Trials).

This kind of regulation is based on *presumed consent*, a type of consent which is founded on a hypothesis based on a scale of values and goods that no one would wish to do without under the rule of law. It is a future-oriented consent, according to which it is reasonable to believe that the person involved would have given his or her consent if given the opportunity to do so and will give that consent once the temporary condition impeding him or her to do so has disappeared (Dworkin 1971). On this kind of consent what matters is not whether consent is realized but whether it is *reasonable to suppose* that the interference at stake would be consented to (Husak 1980). When the action seems to be self-harming and of the sort that most people would choose not to take, there are good, albeit only statistical, grounds for inferring the contrary: 'there are actions of a kind that create a powerful *presumption* that any given actor, if he were in his right mind, would not choose them' (Feinberg 1971, 113). This presumption cannot be aprioristic but must take into account the circumstances surrounding the proposed intervention for it is they that shall determine the measure's legitimacy. These circumstances include 'the individual's competence and rationality, whether or not the intervention affects his or her scale of values, the gravity of the damage that will be avoided, the degree of restriction of liberty, or the use of coercion entailed by the intervention' (Tomás-Valiente 1999, 449). That said, this presumption can be refuted because said presumed consent could be revoked by whoever did not wish to be included in the clinical trials to be carried out in public health crises.

A life-threating illness is a situation in which the future-oriented consent operates because 'it is not in the patient's interest to be denied access to potentially beneficial experimental treatment' (Biros et al. 1995, 1285). The access to experimental treatments 'may be directly beneficial to research subjects, and that enrolling patients in these trials is merely an extension of the physician's obligation to act in the best interest of the patient' (Morrison et al. 2009, 118). In clinical research in emergencies, 'entry of critically ill, unconscious patients into an emergency research protocol without consent has been shown to be not only acceptable to but also appreciated by the patient's families who, in most cases, consented to the patients' continued participation in the protocol' (Biros et al. 1995, 1285).

If this posture is applied to a public health crisis or CBRNE event, we could argue that, in cases in which no effective standard of care is available for a life-threatening illness, participation in a clinical trial may be considered beneficial and such experimental intervention 'may be that which patients would want if they could be fully informed of their expected poor outcome and could indicate their preferences' (Biros et al. 1995, 1284). Persons caught up in a public health crisis of the kind described here may not be able to make capacitated and voluntary decisions to participate in a clinical trial. Participation in the clinical trial will not unbalance the risk-benefit ratio because potential research participants are already at high risk since they have a medical condition with no proven or effective treatment. And, respecting information, even if it can be supplied to obtain consent, it would be still possible to create a presumption that any person, if she were in her right mind, would choose to participate. We could then evaluate the refusal of enrolment as a self-harming action and of the sort that most people would choose not to take, creating a presumption for the participation. However, we need to acknowledge that 'this seems to lead us to a form of paternalism that is so weak and innocuous' (Feinberg 1971, 113).

This normative proposal is in no way intended to undermine the grounds that justify the duty to obtain informed consent in normal circumstances, but only seeks to add some flexibility to its application in such particular cases as the one described here. The general rule has an individual dimension since without the informed, voluntary consent of a competent person, there can be no enrolment; in contrast, the proposal to allow inclusion in clinical trials without consent is collective in so far as it aims to protect all persons from the risks of a life-threatening disease.

10.4 Conclusion

Crisis situations and CBRNE events pose a challenge to the legal and ethical answers we are prepared to give. We need to discuss how clinical trials should be carried out in public health crisis situations and whether participation might be made mandatory. In the context of a global public health crisis, the approach taken here to normative double standards and mandatory participation in designing clinical trials amounts to an appeal to the value of solidarity (see Chap. 11 for a more extensive discussion of this concept in research, and Chap. 9 where one conceptual approach to solidarity is explored in greater depth). Eliminating the double standard would entail a transfer of public goods such that access to the health care technologies enjoyed in developed countries would become available to the most underprivileged, thereby narrowing the breach that exists between the levels of protection existing in different parts of the world for the right to health.

References

Adebamowo, Clement, Oumou Bah-Sow, Fred Binka, Roberto Bruzzone, Arthur Caplan, Jean-François Delfraissy, David Heymann, Peter Horby, Pontiano Kaleebu, Jean-Jacques Muyembe Tamfum, Piero Olliaro, Peter Piot, Abdul Tejan-Cole, Oyewale Tomori, Aissatou Toure, Els Torreele, and John Whitehead. 2014. Randomised controlled trials for Ebola: Practical and ethical issues. *The Lancet* 384: 1423–1424.

Angell, Marcia. 1997. The ethics of clinical research in the third world. *New England Journal of Medicine* 337 (12): 847–849.

Appelbaum, Paul, Loren Roth, and Charles Lidz. 1982. The therapeutic misconception: Informed consent in psychiatric research. *International Journal of Law and Psychiatry* 5: 319–329.

Benatar, Solomon. 2015. Explaining and responding to the Ebola epidemic. *Philosophy, Ethics, and Humanities in Medicine* 10 (5). https://doi.org/10.1186/s13010-015-0027-8.

Biros, Michelle, Roger Lewis, Carin Olson, Jeffrey Runge, Richard Cummins, and Norman Fost. 1995. Informed consent in emergency research. *JAMA* 273 (16): 1283–1287.

Buchanan, David, Sibusiso Sifunda, Nasheen Naidoo, Shamagonam James, and Priscilla Reddy. 2008. Assuring adequate protections in international health research: A principled justification and practical recommendations for the role of community oversight. *Public Health Ethics* 1 (3): 246–257.

Caplan, Arthur, Carolyn Plunkett, and Bruce Levin. 2015. Selecting the right tool for the job. *The American Journal of Bioethics* 15 (4): 4–10.

Caplan, Arthur. 2015. Morality in a time of Ebola. *The Lancet* 385 (9971): e16–e17.

CIOMS (Council for International Organizations of Medical Sciences). 2016. International ethical guidelines for health-related research involving humans. Geneva: CIOMS.

Coleman, Carl. 2009. Vulnerability as a regulatory category in human subject research. *Journal of Law, Medicine and Ethics* 37 (1): 12–18.

Cooper, Ben, Maciej Boni, Wirichada Pan-ngum, Nicholas Day, Peter Horby, Piero Olliaro, Trudie Lang, Nicholas White, Lisa White, and John Whitehead. 2015. Evaluating clinical trial designs for investigational treatments of Ebola virus disease. *PLoS Medicine* 12 (4): 1–14.

Cox, Edward, Luciana Borio, and Robert Temple. 2014. Evaluating Ebola therapies – the case for RCTs. *New England Journal of Medicine* 371 (25): 2350–2351.

Delaney, Martin. 1989. The case for patient access to experimental therapy. *Journal of Infectious Diseases* 159 (3): 416–419.

Diallo, Dapa, Ogobara Doumbo, Christopher Plowe, Thomas Wellems, Ezekiel Emanuel, and Samia Hurst. 2005. Community permission for medical research in developing countries. *Journal of Infectious Diseases* 41 (2): 255–259.

Donovan, Kevin. 2014. Ebola, epidemics, and ethics – What we have learned. *Philosophy, Ethics and Humanities in Medicine* 9 (15): 1–4.

Dworkin, Gerard. 1971. Paternalism. In *Morality and the law*, 107–126. Belmont: Wadsworth.

Edwards, Sarah. 2013. Ethics of clinical science in a public health emergency: Drug discovery at the bedside. *The American Journal of Bioethics* 13 (9): 3–14.

Emanuel, Ezekiel, David Wendler, and Christine Grady. 2000. What makes clinical research ethical? *JAMA* 283 (20): 2701–2711.

Emanuel, Ezekiel, David Wendler, Jack Killen, and Christine Grady. 2004. What makes clinical research in developing countries ethical: The benchmarks of ethical research. *The Journal of Infectious Diseases* 189 (5): 930–937.

European Union. 2014. *Regulation (EU) 536/2014, of the European Parliament and of the Council of 16 April 2014 on Clinical Trials on Medicinal Products for Human Use, and Repealing Directive 2001/20/EC.*

Ezeome, Emmanuel, and Christian Simon. 2010. Ethical problems in conducting research in acute epidemics. *Developing World Bioethics* 10 (1): 1–10.

Feinberg, Joel. 1971. Legal paternalism. *Canadian Journal of Philosophy* 1 (1): 105–124.

Gelinas, Luke, Alan Wertheimer, and Franklin Miller. 2016. When and why is research without consent permissible? *Hasting Center Report* 46 (2): 35–43.
Goodman, Jesse. 2014. Studying secret serums – Towards safe, effective Ebola treatments. *New England Journal of Medicine* 371 (12): 1086–1089.
Hantel, Andrew, and Olopade Christopher Olusola. 2015. Drug and vaccine access in the Ebola epidemic: Advising caution in compassionate use. *Annals of Internal Medicine* 162 (2): 141–142.
Hodge, James, Leila Barraza, Gregory Measer, and Asha Agrawal. 2014. Global emergency legal response to the 2014 Ebola outbreak. *Journal of Law, Medicine and Ethics* 42 (4): 595–601.
Husak, Douglas. 1980. Paternalism and autonomy. *Philosophy and Public Affairs* 10 (1): 29–46.
Joffe, Steven. 2014. Evaluating novel therapies during the Ebola epidemic. *JAMA* 312 (13): 1299–1300.
Larkin, Megan. 2015. Defining compensable injury in biomedical research. *Health Matrix: The Journal of Law-Medicine* 25: 309–382.
London, Alex John. 2000. The ambiguity and the exigency: Clarifying 'standard of care' arguments in international research. *Journal of Medicine and Philosophy* 25 (4): 379–397.
———. 2009. Clinical research in a public health crisis: The integrative approach to managing uncertainty and mitigating conflict. *Seton Hall Law Review* 39 (4): 1173–1202.
Lurie, Nicole, Teri Manolio, Amy Patterson, Francis Collins, and Thomas Frieden. 2013. Research as part of public health emergency response. *New England Journal of Medicine* 368 (13): 1251–1255.
Lurie, Peter, and Sidney Wolfe. 1997. Unethical trials of interventions to reduce perinatal transmission of the human Immunodeficiency virus in developing countries. *New England Journal of Medicine* 337 (12): 853–856.
Macklin, Ruth. 1999. *Against relativism. Cultural diversity and the search for ethical universals in medicine*. New York: Oxford University Press.
———. 2004. *Double standards in medical research in developing countries*. New York: Cambridge University Press.
McConnell, Terrance. 2010. The inalienable right to withdraw from research. *Journal of Law, Medicine & Ethics* 38: 840–846.
Meier, Benjamin Mason, and Larisa Mori. 2005–2006. The highest attainable standard: Advancing a collective human right to public health. *Columbia Human Rights Law Review* 37: 101–147.
Miguel A. Ramiro Avilés. 2014. Paving the road to negligence: The compensation for research-related injuries in Spain. *Accountability in Research* 22 (2):106–119.
Miller, Franklin, and Howard Brody. 2003. A Critique of clinical equipoise. Therapeutic misconception in the ethics of clinical trials. *The Hastings Center Report* 33 (3): 19–28.
Morin, Karin, Herbert Rakatansky, Frank Riddick, Leonard Morse, John O'Bannon, Michael Goldrich, Priscilla Ray, Matthew Weiss, Robert Sade, and Monique Spillman. 2002. Managing conflicts of interest in the conduct of clinical trials. *JAMA* 287 (1): 78–84.
Morrison, Anne, Irwin Horwitz, and Matthew Carrick. 2009. Ethical and legal issues in emergency research. *Journal of Surgical Research* 157 (1): 115–122.
O'Mathúna, Dónal. 2015. Research ethics in the context of humanitarian emergencies. *Journal of Evidence-Based Medicine* 8 (1): 31–35.
Presidential Commission for the Study of Bioethical Issues. 2015. *Ethics and Ebola. Public health planning and response*. Washington, DC: Presidential Commission for the Study of Bioethical Issues.
Ramiro Avilés, M.A. 2015. Paving the road to negligence: The compensation for research-related injuries in Spain. *Accountability in Research* 22: 106–119.
Rid, Annette. 2015. The goals of research during an epidemic. *The American Journal of Bioethics* 15 (4): 47–50.
Rid, Annette, and David Wendler. 2011. A framework for risk-benefit evaluations in biomedical research. *Kennedy Institute of Ethics Journal* 21 (2): 141–179.
Rid, Annette, and Ezekiel Emanuel. 2014. Ethical considerations of experimental interventions in the Ebola outbreak. *The Lancet* 384 (9957): 1896–1869.

Rothstein, Mark. 2015. The moral challenge of Ebola. *American Journal of Public Health* 105 (1): 6–8.
Schüklenk, Udo. 2014. Access to unapproved medical intervention in cases of catastrophic illness. *American Journal of Bioethics* 14 (11): 20–22.
Shah, Seema, David Wendler, and Marion Danis. 2015. Examining the ethics of clinical use of unproven interventions outside of clinical trials during the Ebola epidemic. *The American Journal of Bioethics* 15 (4): 11–16.
Shaw, David. 2014. Randomization is essential in Ebola drug trials. *The Lancet* 384: 1667.
Smith, Maxwell, and Diego Silva. 2015. Ethics for pandemics beyond influenza: Ebola, drug-resistant tuberculosis, and anticipating future ethical challenges in pandemic preparedness and response. *Monash Bioethics Review* 33 (2–3): 130–147.
Strauss, Stephen. 2014. Ebola research fuelled by bioterrorism threat. *Canadian Medical Association Journal* 186 (16): 1206.
Thompson, Dennis. 1987. *Political ethics and public office*. Cambridge: Harvard University Press.
Tomás-Valiente, Carmen. 1999. The justification of paternalism. *Rechtstheorie* 30: 431–460.
WHO (World Health Organization). 2014. *Ethical issues related to study design for trials on therapeutics for Ebola virus disease*. Geneva: WHO.
Yearby, Ruqaiijah. 2003–2004. Good enough to use for research, but not good enough to benefit from the results of that research: Are the clinical HIV vaccine trials in Africa unjust. *De Paul Law Review* 53: 1127–1153.

Chapter 11
A Solidarity-Based Framework for Ethical Clinical Research in CBRNE

Oren Asman and Yechiel Michael Barilan

Abstract Ordinary standards of research ethics do not fit the circumstances of CBRNE. This chapter argues that the value of solidarity may sustain certain moral principles for research in CBRNE, and that standards for research ethics adopted to other special circumstances may be imported into CBRNE research. The key adaptation is the model of advance directives to research which may be accepted by people who are likely to be victims of CBRNE, such as workers in emergency services and high-risk industry (e.g. nuclear plants).

Keywords Clinical research · CBRNE · Clinical ethics · Solidarity · Bioethics · Disasters ethics

11.1 Overview

Chemical, biological, radiological, nuclear and explosive (CBRNE) events are unusual, rare events, typically involving mass casualties. The most prepared for and feared of such events, have never even happened Hopefully, they never will. By their own nature - events of low probability of occurrence, yet with very high risk in terms of impact – CBRNE pose significant hurdles to the execution of research and informed consent to clinical research both during and in the aftermath of such events. The ethics of any kind of research in disasters is quite challenging. This challenge extends even to merely observational and interview studies (Barron Ausbrooks et al. 2009).

O. Asman (✉)
Faculty of Medicine, Tel Aviv University, Tel Aviv, Israel

Sechenov First Moscow State Medical University, Moscow, Russia
e-mail: asman@tauex.tau.ac.il

Y. M. Barilan
Faculty of Medicine, Tel Aviv University, Tel Aviv, Israel

D. P. O'Mathúna, I. de Miguel Beriain (eds.), *Ethics and Law for Chemical, Biological, Radiological, Nuclear & Explosive Crises*, The International Library of Ethics, Law and Technology 20, https://doi.org/10.1007/978-3-030-11977-5_11

This chapter offers an argument supporting a duty to conduct a specific kind of clinical research in CBRNE, indicate circumstances allowing for such research to deviate and differ from routine standards of ethical research, and offer ways for doing so, mainly the adaptation of existing instruments that facilitate clinical research in challenging circumstances. This chapter aims to highlight features of research ethics in CBRNE that differentiate it from research ethics in other large-scale disasters (e.g. Daugherty and White 2010; O'Mathúna 2012).

The extreme and disruptive nature of CBRNE may coax a perception of supreme emergency that disrupts ordinary ethical standards. In this chapter we argue that CBRNE related ethical problems are amenable to examination by established bioethical values and by adaptation of existing solutions to non-CBRNE research situations. Some standard ethical principles for clinical research are elaborated in more detail in Chap. 10 of this volume. Our working hypothesis for this chapter is that meticulous patchwork re-design may be morally preferable to policies of exception and are likely to attract higher rates of legitimization.

11.2 CBRNE as a Construct of Exception

There is broad consensus that CBRNE events are extraordinary occurrences that require special administrative and clinical policies and preparedness. Because these policies must be evidence-based and scientifically informed, they need support by research. This need brings to the fore the challenge of informed consent to clinical research in CBRNE. At first glance, one may have the impression that, in relation to research ethics, CBRNE research is like research in other medical emergencies such as pertaining to life-saving procedures in the emergency department. However, significant differences call for our attention. First, whereas in the emergency room it is usually the patient who is suffering from the medical and psychological shock of the emergency, in CBRNE both the caring environment and the patient could be seriously afflicted and their capacity for sound decision making and for oversight over their execution are stretched to the limits. Second, whereas the average person knows something about "a heart attack", "acute bleeding", "motor accidents" and other ordinary personal emergencies, much less common knowledge exists regarding CBRNE. Third, the very notion of public emergency is easily associated with the ethics of war, especially the "war on terror". In the eyes of many, CBRNE is not merely an "emergency" akin to the "supreme emergency", a term originally coined by Michael Walzer in relation to the critical moments of the Second World War (Coady 2004). Few think that a CBRNE event is likely to destroy the free world, not even a single country; nonetheless, the human and social costs of CBRNE, and of badly managed CBRNE, seem to be devastating.

This ominous representation seems to provide a convincing argument in defense of a set of ethical standards in CBRNE research ethics. Because, no scientific medicine is possible without research, and society is not going to forgo scientific medical preparedness and responses to CBRNE, research must take place. Because in

CBRNE, ordinary standards of research ethics are either impractical or meaningless, and because dissolution of all ethical standards is unheard of, an alternative set must prevail. One example, that of double standards for clinical research in different cultures, is examined in Chap. 10.

However, this reasoning is far from satisfactory. Justification for deviation from one set of standards is a far cry from arguing for an alternative mode of regulation. We must also keep in mind, that deviation from standards may entail *stricter* ones. The argument for research ethics in CBRNE is not an argument in favor of generalized relaxation of ethical and legal requirements. In this chapter, we explicate an outline for informed consent to research in CBRNE. We suggest solidarity as a core value of this outline and argue that even if specific applications of informed consent are substituted by others, new and bold ones, the doctrine of consent should be preserved in a democracy.

11.3 Solidarity, Liberal Democracy and Bioethically-Laden Health Related Policy

The first English language uses of "solidarity" are traceable to the mid nineteenth century as related to group responsibility. The contemporary academic discourse of "solidarity" is traceable to Emile Durkheim. In his 1893 doctoral dissertation and first major work – "The Division of Labor in Society" – Durkheim presented solidarity as the abstract foundation of social cohesion and saw it as a descriptive term referring to a fundamental social good.(Durkheim, 1984)

Hannah Arendt contraposed solidarity with pity. Whereas pity motivates localized responses to unfortunate occurrences, solidarity motivates the creation of long-term social bonds whose ultimate underpinnings is human dignity, not human misery and folly. A drunk man falling in a ditch may elicit mere pity. However, even though solidary may be roused by emotion, it is guided by rationality (Arendt 1963). In this line, if we are to view CBRNE as a foreseeable accident of our shared social life and culture, society's rational preparedness and response to CBRNE should rest on solidarity.

The first normative meaning of "solidarity" is an implicit commitment to be morally obliged by the consequences of decisions (e.g. the law of the state) or actions (e.g. collective guilt) made by a group to which a person belongs to (Mason 1998). This conceptualization of solidarity is relevant to joint decision making in issues such as foreign relations and environmental policy. In contemporary democracies, a group cannot impose on others decisions of a personal nature, such as the acceptance of medical care or participation in medical experimentation. It follows that while preparedness and response to CBRNE is a matter of joint decision making (the first conceptualization of solidarity), it depends on clinical research whose governance calls for a richer rendering of solidarity, one that interacts with individual human rights and the internal morality of medicine and science.

A second normative conceptualization of "solidarity" is about a primary moral motivation. It is a reason to make certain choices and to act in certain ways in response to moral needs. This last construct of solidarity is found among the core set of bioethical principles (along with justice, respect for dignity, respect for autonomy, beneficence, non-maleficence, and vulnerability). In contemporary applied ethics, "solidarity" has been developing as a fundamental social *norm*, which is forward looking (i.e. future action) rather than backward looking (e.g. collective guild and desert). Thalos (2012) offers a motivational conception for solidarity. Below we explicate this emerging subtype of "solidarity".

Solidary is a perception of moral commitment towards some people based on a morally relevant shared trait or interest (or a set thereof). Solidarities and claims to solidarity-related behaviors might combine and clash with each other. From the perspective of the individual, solidarity is a matter of personal choice in relation to incomplete duties (the recipient is known, such as a needy child; but there is nobody designated as accountable for action). It is often alleged that with modernism and globalization, solidarity has been weakening (Ten Have 2016). However, it might also be possible that an ever-growing awareness of diverse and apparently conflicting, solidarities is responsible for this perception. It is also possible that solidarity's role in social cohesion has diminished relative to other values, such as liberty and bureaucratization of society, without adversely impacting the moral calling of solidarity and solidarity-motivated policies.

Talking about solidarity as a value in public policies, we need to explicate its nature, its standing relative to other solidarities and relative to other values such as respect for personal autonomy. In healthcare ethics, solidarity is focused on health related human vulnerabilities.

The bioethical conceptualization of solidarity raises four key questions: the first is the identification of basic values that are related to healthcare and the human condition as such; the second is the relationship between solidarity and beneficence. The cement of beneficence allows solidarity to occur (Loewy 1991); the third is the relationship between solidarity and respect for personal autonomy. Solidarity promotes uniform acts (e.g. vaccination, blood donations, alms giving) to the benefit of the needy; respect for autonomy promotes individual choice, even in disregard of the needy. Fourth, sometimes public policy needs to balance one kind of solidarity against the other. For example, solidarity with victims of sexual abuse and solidarity with babies may push in opposing directions. This is one reason why solidarity is a philosophically raw concept that depends on explication within a broader normative matrix. In the following paragraphs, we explicate the value of solidarity in bioethics, sometimes departing a little from the last major effort in this direction (Prainsack and Buyx 2011).

Bioethics is the moral ethos regarding health and healthcare in the democratic society. Hence, the relevant solidarity is only health and health-care related, and the most relevant bioethical value in this regard is human vulnerability. Bioethics has also been emphasizing environmental awareness and concerns for future generations. While the latter aspect has little direct bearing on medical ethics, it may resurface in relation to CBRNE, as in a combined clinical and environmental crisis.

Solidarity creates a justification for spending public moneys, burdening everybody with fungible costs. Even legal structures that tightly restrain the power of the government to spend (e.g. the Constitution of the USA), have accepted the notion that the public should assist victims of disasters and vulnerable people with life-saving needs. Some policies are explicable in terms of the mutual benefit of insurance and similar schemes of joint assistance. However, we invoke the value of solidarity even in the absence of risk-benefit calculations.

Solidarity creates defaults. Solidarity-based policies do not trump personal choice over body and person (Barilan 2012). The burden they place on individuals is minor and often not beyond the fungible. It follows that with organ donation from the brain dead, solidarity may justify opting-out schemes, but not compulsory ones. Solidarity may also justify laws constraining the sales of unhealthy food. Thus, society promotes solidarity-oriented choices, creating incentives and fostering publicity campaigns. Whereas, it is possible to behold the encouragement of "healthy" lifestyle as "nudges" in the benefit of the common good, other policies are clearly in the benefit of the few.[1] Laws obliging wheelchair accessibility are one example. Liberal societies have found acceptable the burden of making businesses accessible to wheelchairs, but not the burden of medical research without informed consent. This is because we do not behold research without consent as a mere burden, but offensive to human dignity.

In sum, solidarity in bioethics is about justification of defaults, and of pecuniary burdens in the benefit of health and healthcare and the protection of the environment. The pre-bioethical conceptions of solidarity – ties of social cohesion, specific shared traits or interests among certain groups of people, and mutual responsibility for joint decision-making - might also contribute to solidarity-informed research ethics in CBRNE.

11.4 Society and Disaster Preparedness

The term "CBRNE" covers numerous and diverse scenarios. Some might pose significant risk to the social fabric and to a large number of human lives. The likelihood that a large-scale CBRNE takes place is unclear. However, the argument from solidarity also supports policies of disaster preparedness that are not subjected to cost-effectiveness criteria only. Even though it might make sense not to prepare for and strive to improve the standards of care for CBRNE victims, solidarity calls for some investment in coping with risks that are very highly groups-specific. Even though only a fraction of the population lives in the vicinity of nuclear plants, these plants exists in the name and benefit of society at large. It is also true that very many people object to these plants and go to considerable length to have them located as remotely as possible (The NIMBY Phenomenon; see Hermansson 2007). All the

[1] Indeed, one does not find "solidarity" in the influential book on "nudges" (Sunstein and Thaler 2009).

while, precisely because these activities and plants are legal, solidarity in the sense of shared responsibility for social choice is a relevant value in relation to an organized response to CBRNE. Not only are CBRNE products (unintentional albeit) of social policy and action, CBRNE are also loci of social vulnerability to highly intensive technological capacity. CBRNE cannot evolve from small scale private operations, but almost always from state and corporate level of management and responsibility. It follows that CBRNE are products of large scale social operations and institution. They call for preparedness and response at a large social scale. Society offers a certain level of protection to individual victims, regardless of many considerations of cost. One reason is the solidarity with victims; the other reason is the preservation of social fabric, which might be badly affected by abandonment of even a few members of society. This second consideration bears out the older meaning of "solidarity".

Michael Tilly maps out human societies on a Cartesian matrix comprised of axes of democracy and governance (Tilly 2003). Societies that are low in democracy and low in governance are "failed states". They offer neither liberties nor services. Societies that are high in democracy and low in governance are of the libertarian kind. Societies that rank highly in governance but low in democracy are tyrannical. These societies show high levels of control and state power, but little of it serves the needs of the people. The upper right corner reflects an ideal construct of "effective citizenship" in a functioning democracy. Liberties are protected and basic services are available for the people (These services may be available through the private markets. Governance is indirect, as society regulates the markets). People are effective citizens because they can bring their liberties to bear on social institutions, goods and their governance. Figure 11.1 renders this scheme, showing also how immediate CBRNE responses push society away from governance *and* from

Fig. 11.1 The influence of CBRNE crises on democracy and governability

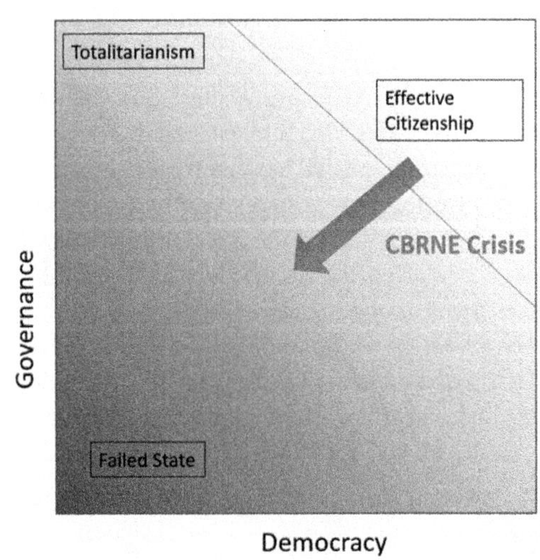

democracy. CBRNE events challenge solidarity as referring to social cohesion and functioning.

An appropriate social response to CBRNE must reflect the two fundamental goods – the promotion of the common good, and protection of people from the subjection of anybody's interests to either the interests or will of others, without an appropriate structure of consent (Shapiro 2003).

It is crucial to realize that even in emergencies, both values need to be respected. Even though it may be unrealistic to process informed consent to research during CBRNE, it is possible to plan in advance, educate people and find alternative modes of regulation that aim at the preservation of people's liberties, interests and dignity. Key to such modes of regulation is the understanding that clinical research is one of the services (part of "governance") society strives to offer to its needy members, and the understandings that exemptions from ordinary informed consent do not entail exemption from the application of the best relevant instrument of consent and from consent-based governance. Extreme CBRNE pose threats to the very fabric of society. But this in itself may not justify loosening of ethical standards, as temporary sacrifices of individual rights for the sake of society. The incorporation of needed research in CBRNE response is in itself a restorative act of the governance axis of society.

11.5 An Outline for Solidarity-Informed Research Ethics in CBRNE

In light of the above considerations, one may delineate some principles for research in CBRNE. In the first step, we focus on groups of workers: those employed in the relevant disaster industry (e.g. a nuclear plant), along with rescue teams and healthcare professionals. These people must be educated in advance with regard to CBRNE preparedness. Part of this education may cover issues of research ethics. Workers who choose not to participate may opt out at this stage.

CBRNE preparedness may contain research modules, research protocols prepared in advance for situations of clinical equipoise, specific clinical questions to which there is not a standardized, evidence-based answer, and to which there exists more than one possible intervention. These research protocols may be reviewed by a research ethics review board (IRB or a special Ethics Review board set up for CBRNE events) allowing matters to be considered without the urgency of an ongoing CBRNE event. Such protocols may be somewhat general in nature, but allow in-depth examination that may not be possible in real time.

Advance preparation may facilitate not only good clinical practice in actual research, but will increase the transparency and resolution of information regarding the planned research and its legitimization. All of this may be incorporated into the education phase in disaster preparedness for the relevant workers. The World Health Organization (WHO) has already research packages prepared for disaster research.

It is possible to inform the public at large about such research plans (e.g. have the information posted on the internet). However, there are good reasons to assume that the vast majority of people will not find the issue salient and will not engage with the topic at the intensity and seriousness necessary for the consolidation of consent. The targeted groups of workers, on the other hand, have very good reasons to be concerned about CBRNE. This is their specific group solidarity, a layer of solidarity that comes in addition to their mere participation in society. The rational and social (i.e. "political nature") of human beings combine need with relevant rational valuation.

Governance by consent and solidarity-based democracy entail the duty to inform the participants and the public post factum. The participants (as well as victims chosen not to be included in the research) have a right to know, and the authorities are accountable before them, as well as before the public at large. Because CBRNE may justify exemptions from ordinary informed consent to research, it does not create waivers of rights to know and to ownership. See Chap. 10 for further discussion of mandatory participation in clinical trials. Rather, within a framework of solidarity, one may expect equal public participation in the sharing of knowledge, oversight, and benefits of CBRNE research. The information and benefits from such research should be within the public domain, not owned by private bodies. Because of their personal involvement and the probable implications for their future health, participants have a special standing in relation to the results of research.

When a victim is enrolled in clinical research whose both arms are in clinical equipoise, care should be identical to the care he or she would receive without participation in research. This is a good guarantee that participation in research does not compromise clinical standards of care. Some research protocols may require acceptance of burdens, such as medical exams and monitoring that are dedicated to the proper comparisons of the two equipoise-related modalities of care. Is it ethical to expose non-consenting victims to this kind of burden?

The simplest justification for equipoise-related research is derived from the fact that care during research is identical to care given outside of research. The only difference is the additional monitoring, collection and processing of data. Quality control is integral to good care and offers an additional weight of justification to the monitoring, collection and processing of data. Is it possible to justify the burden of research related monitoring?

Lack of research means practice without proper scientific evidence and without an attempt to improve the knowledge base. It may be argued that such care is morally inferior to the provision of the same kind of care (here is the equipoise element) under the scrutiny of research. There is also a sound counter-argument, saying that it is more respectful of persons to let them be (natural arbitrariness) than have them subjected to some kind of deliberate scrutiny and manipulation by others, benevolence notwithstanding. Hence, it may be argued that whenever we have good reasons to suppose that one treatment option is *inferior and harmful*, and that could only be examined as a part of research within CBRNE settings, and that this research must be accompanied by additional monitoring is likely to either confirm or reject the inferiority hypothesis, such research is justified. It is in the benefit and dignity of both victims and the public that harmful modalities of care do not die hard.

Within the context of this chapter, even a weak argument would suffice. This argument would be, that a selective, solidarity-relevant group of properly informed workers may be enrolled in equipoise-relevant research in circumstances specific to CBRNE. With an opting out option (in advance and in real time, when possible), the mere *reasonableness* of the superiority of research over non-research-based care is morally sufficient.

The argument may become stronger upon the incorporation of instruments that already help people cope with anticipated lack of full capacity to consent to participation in research. One such instrument is advanced directive consenting to research on dementia patients (Andorno 2016). The most appropriate timing for signing such a document when that patient is already suffering from the relevant condition or conditions, and yet he or she is fully competent to weigh the alternatives and make an informed choice. The young and healthy may not dedicate the appropriate solemnity to the question in hand; those whose mental faculties are already deteriorating may not be able to make a choice.

In a similar vein, the three target groups designated by this chapter – workers in CBRNE-related plants, healthcare responders and security providers should be both aware of the CBRNE risks and capable of consenting in advance to participation in certain clinical research protocols in case a disaster occurs. These workers consent to sharing some private data, and especially medical information, with their employers. They undergo close monitoring and, in many places, submit urine samples for drug testing. It seems reasonable to suppose that they can also give consent in advance to the monitoring and data processing associated with research on equipoise-related questions, whenever such research does not add medical exams and procedures not requested outside of research. Work-related consent to sharing information with employers may not be free of ethical issues; however, the ethics and law of consensual exposure of private data by employees in the emergency, security and health sector, is one existing instrument which may serve as a legal and conceptual anchor for CBRNE-related research.

Another instrument is the waiver of informed consent in clinical research in emergencies. Many emergency department patients suffer from sudden, unexpected catastrophes that require speedy and bold interventions (Scott et al. 2011). One recent example is thrombolytic therapy to stroke patients. This is a risky treatment that has revolutionized stroke care owing to clinical trials conducted on patients within minutes of presentation in emergency departments. Tight regulation and strictly-selected participants and interventions allowed the relaxation of informed consent standards without which new stroke therapies would not have materialized. The analogy between clinical research on personal medical emergencies and clinical research in social and clinical emergencies is only partial. However, it is close enough to justify the importation of some principles and precautions taken in the former to the planning of the latter.

11.6 Conclusion

We have presented solidarity as a foundation for an argument for the duty to conduct clinical research in CBRNE care, at least on equipoise-related questions, and whenever there is good scientific grounds to suppose that a prevailing practice is actually harmful, or less beneficial. A weaker argument supported clinical research whenever there are good reasons to suppose that a novel intervention might become a new, superior standard of care relative to the prevailing modes of care. Chap. 10 provides other approaches to these ethical challenges with CBRNE research.

We suggested having openly prepared research protocols as part of CBRNE preparedness, especially regarding the care of CBRNE related workers, emergency and health services. Solidarity creates the motivation and justification for CBRNE research, especially research that deviates from ordinary standards of informed consent. Solidarity should also allow non-willing victims to opt out of research. Solidarity-based research requires transparency, public oversight and public ownership and benefits from research, with special duties of accountability and continuous care for those enrolled in the research.

CBRNE are extraordinarily rare events. However, existing and emerging instruments of research ethics, such as waiver to informed consent, advanced directives to research, review and approval of outline protocols for future event-related research, and possibly the appointment and activation of emergency research-ethics review boards may be suitable for CBRNE, and may provide continuity with ordinary care and clinical research ethics.

References

Andorno, R. 2016. Integrating advance research directives into the European legal framework. *European Journal of Health Law* 23: 158–173.

Arendt, H. 1963. *On revolution*. New York: Viking.

Barilan, Y.M. 2012. *Human dignity, human rights and responsibility: The new language of bioethics and biolaw*. Cambridge, MA: MIT Press.

Barron Ausbrooks, C.Y., E.J. Barrett, and M. Martinez-Cosio. 2009. Ethical issues in disaster research: Lessons from Hurricane Katrina. *Population Research and Policy Review* 28 (1): 93–106.

Coady, C.A.J. 2004. Terrorism, morality, and supreme emergency. *Ethics* 114: 772–789.

Daugherty, E.L., and D.B. White. 2010. Conducting clinical research during disasters. *Virtual Mentor* 12 (9): 701–705.

Durkheim, E. 1984. *The division of labor in society*. New York: The Free Press.

Hermansson, H. 2007. The ethics of NIMBY conflicts. *Ethical Theory and Moral Practice* 10: 23–34.

Loewy, Erich H. 1991. *Suffering and the beneficent community: beyond libertarianism*, 80. Albany: State University of New York Press.

Mason, A. 1998. *Solidarity*. Routledge encyclopedia of philosophy. https://www.rep.routledge.com/articles/thematic/solidarity/v-1. Accessed 8 Oct 2018.

O'Mathúna, D. 2012. Roles and challenges for IRBs with disaster research. *Research Practitioner* 13 (5): 167–174.

Prainsack, B., and A. Buyx. 2011. *Solidarity: Reflections on an emerging concept in bioethics*. London: Nuffield Foundation. http://nuffieldbioethics.org/wp-content/uploads/2014/07/Solidarity_report_FINAL.pdf. Accessed 8 Oct 2018.

Scott, E., et al. 2011. Community views on neurologic emergency treatment trials. *Annals of Emergency Medicine* 57: 346–354.

Shapiro, I. 2003. *The state of democratic theory*. Princeton: Princeton University Press.

Sunstein, C.R., and R.H. Thaler. 2009. *Nudge: Improving decisions about health, wealth and happiness*. London: Penguin.

Ten Have, Henk, ed. 2016. *Encyclopedia of global bioethics*, 216–220. Cham: Springer International Publishing.

Thalos, M. 2012. Solidarity: A motivational conception. *Philosophical Papers* 41: 57–95.

Tilly, M. 2003. *The politics of collective violence*. Cambridge: Cambridge University Press.

Chapter 12
The Current Legal Framework on Data Protection in CBRNE Crises: A General Exposition

Asier Urruela Mora

Abstract Due to the fact that CBRNE situations are not a major issue when regulating data processing (both at international and internal level) only by means of the exceptions established by the general regulation on processing of personal data it is possible to achieve the setting of a legal framework on this topic in CBRNE contexts. The main purpose of the present paper is to point to the existing regulation at this level in the European context (including regulation produced by the Council of Europe and the European Union legal framework) and to highlight the possible need for legal reforms so as to avoid a lack of protection in CBRNE situations.

Keywords Personal data protection · CBRNE crises · EU legal framework on personal data protection · Convention 108 of the Council of Europe · Threats to public security

12.1 Introduction

Accidental or deliberate Chemical, Biological, Radiological, Nuclear and Explosive (CBRNE) major crises highlight some important aspects related to the field of data protection and privacy issues. Under those exceptional circumstances, data are likely to be collected without the data subjects' consent, under highly stressful conditions, in the absence of normal infrastructure, and in the midst of political and legal uncertainty. Moreover, it is perfectly possible to imagine a situation when the authorities responsible for the life and health of an entire population might need to gain access to databases which are usually highly protected in order to mitigate the consequences of a CBRNE major crisis event: this could be the case, for instance, with bioterrorist attacks or sabotage of nuclear resources. In any case, we have to point out from the beginning of our study that CBRNE situations are not a major

A. Urruela Mora (✉)
University of Zaragoza, Zaragoza, Spain
e-mail: asier@unizar.es

D. P. O'Mathúna, I. de Miguel Beriain (eds.), *Ethics and Law for Chemical, Biological, Radiological, Nuclear & Explosive Crises*, The International Library of Ethics, Law and Technology 20, https://doi.org/10.1007/978-3-030-11977-5_12

issue when regulating fields like data processing (at European Union (EU), international and internal levels). Consequently only by means of the exceptions established by the general regulation on processing of personal data is it possible to achieve the setting of a legal framework related to the processing of personal data in the CBRNE contexts.

The main purpose of the present paper is to point to the existing regulation at this level in the European context and to highlight the possible need for legal reforms so as to avoid a lack of protection in CBRNE situations.

To consider these points, the chapter is structured around two basic milestones:

1. The relevance of regulation issued from the Council of Europe. The most specific convention in this regards is the Convention for the Protection of Individuals with regard to Automatic Processing of Personal Data of 1981.
2. Analysis of the regulation existing at EU level. This involves mainly the Charter of Fundamental Rights of the European Union and Article 16 of the Treaty on the Functioning of the European Union. The latter article recognises the right of each person to the protection of personal data concerning him or her and mandates the European Parliament and the Council to lay down rules relating to the protection of natural persons with regard to the processing of personal data by Union institutions, bodies, offices and agencies, and by the Member States when carrying out activities which fall within the scope of Union law, and the rules relating to the free movement of such data. The following are the applicable legislation at EU level:

   - Regulation 2016/679 of the European Parliament and of the Council on the protection of natural persons with regard to the processing of personal data and on the free movement of such data, and repealing Directive 95/46/EC (General Data Protection Regulation) that applies from 25 May 2018.
   - Directive (EU) 2016/680 of the European Parliament and of the Council of 27 April 2016 on the protection of natural persons with regard to the processing of personal data by competent authorities for the purposes of the prevention, investigation, detection or prosecution of criminal offences or the execution of criminal penalties, and on the free movement of such data , and repealing Council Framework Decision 2008/977/JHA. The analysis in this chapter will focus on the aspects related to the processing of personal data (and basically the limitation thereof) in the context of CBRNE crises which is a subject directly connected with the scope and the subject-matter of Directive 2016/680. Nevertheless, it is relevant to point out that Directive 95/46/EC of the European Parliament and of the Council of 24 October 1995 on the protection of individuals with regard to the processing of personal data and on the free movement of such data and other relevant documents are analysed elsewhere in this volume in Chap. 13.[1] This Directive continutes to provide the legal principles and framework for the existing data protection regulations in the EU.

[1] We will also leave outside the scope of our analysis the issues connected with data protection and privacy in the communication sector regulated by Directive 2002/58/EC of the European Parliament

12.2 The Regulation Produced by the Council of Europe

12.2.1 Introduction

The Council of Europe is not a part of the European Union architecture. In fact, it comprises a wider range of States than the ones belonging to the EU and is considered to be the continent's leading human rights organisation.[2] However, insofar as all Member States are also part of the Council of Europe, its conventions and resolutions are particularly important in the EU context.

In the concrete arena of data protection and privacy issues, the most remarkable legal tool produced by the Council of Europe is the Convention for the Protection of Individuals with regard to Automatic Processing of Personal Data of 1981. Additionally, the Convention for the protection of Human Rights and Dignity of the Human Being with regard to the Application of Biology and Medicine (Convention on Human Rights and Biomedicine) of 1997 has also to be mentioned when considering the necessity of interventions in human beings in the health field (particularly, its Article 10).[3]

12.2.2 The Convention for the Protection of Individuals with Regard to Automatic Processing of Personal Data of 1981

The Convention for the Protection of Individuals with regard to Automatic Processing of Personal Data of 1981, usually called Convention 108, was drawn up within the Council of Europe by a committee of governmental experts under the authority of the European Committee on Legal Co-operation (CDCJ), and opened for signature by the Member States of the Council of Europe on 28 January 1981 in Strasbourg.[4]

---

and of the Council of 12 July 2002 (Directive on privacy and electronic communications) amended by Directive 2006/24/EC and by Directive 2009/136/EC.

[2] Council of Europe. The Council of Europe in brief. http://www.coe.int/en/web/about-us/who-we-are. Accessed 30 Nov 2016.

[3] Article 10 (Private life and right to information) Convention on Human Rights and Biomedicine: "*1. Everyone has the right to respect for private life in relation to information about his or her health. 2. Everyone is entitled to know any information collected about his or her health. However, the wishes of individuals not to be so informed shall be observed. 3. In exceptional cases, restrictions may be placed by law on the exercise of the rights contained in paragraph 2 in the interests of the patient*".

[4] Nevertheless, recent Court rulings (both from the Court of Justice of the European Union and from the European Court of Human Rights) reaffirming the need for a strong protection of individuals with regard to the processing of personal data have bring to light the necessity of the modernisation of Convention 108 and its global promotion. Consequently, the Council of Europe

Our main interest in this Convention relies on the fact that it was the first legally binding international instrument adopted in the field of data protection. Indeed, nowadays it still remains the only binding international legal instrument with a worldwide scope of application in the field of data privacy, open to the signature of any country, including all those which are not Members of the Council of Europe. As stated in its article 1, its purpose was "*to secure in the territory of each Party for every individual, whatever his nationality or residence, respect for his rights and fundamental freedoms, and in particular his right to privacy, with regard to automatic processing of personal data relating to him (data protection)*". In doing so, the Convention sets out a whole range of minimum standards aimed at protecting individuals against abuses which may accompany the collection and processing of personal data.

In its chapter II the Convention establishes the basic principles for data protection and, in particular, Article 5 refers to quality of data, which includes:

(a) Fair and lawful obtaining and processing.
(b) Storage for specified and legitimate purposes and not used in a way incompatible with those purposes.
(c) Personal data undergoing automatic processing shall be adequate, relevant and not excessive in relation to the purposes for which they are stored.
(d) Personal data undergoing automatic processing shall be accurate and, where necessary, kept up to date.
(e) Personal data undergoing automatic processing shall be preserved in a form which permits identification of the data subjects no longer than is required for the purpose for which those data are stored.

Related to special categories of data, article 6 states that personal data revealing racial origin, political opinions or religious or other beliefs, as well as personal data concerning health or sexual life, may not be processed automatically unless domestic law provides appropriate safeguards. The same shall apply to personal data relating to criminal convictions.

Additional safeguards for the data subject are set in article 8 which establishes that any person shall be enabled:

(a) to establish the existence of an automated personal data file, its main purposes, as well as the identity and habitual residence or principal place of business of the controller of the file;

is updating its Personal Data Protection Convention – "Convention 108" – with two key aims: on one side, addressing challenges for privacy resulting from the use of new information and communication technologies; secondly, strengthening the convention's follow-up mechanism. The modernisation process also aims at bringing together the various normative frameworks that have developed in different regions of the world and provide a multilateral framework that is flexible, transparent and robust, facilitating the flow of data across borders while providing effective safeguards against abuse. See Council of Europe. Modernisation of the Data Protection "Convention 108". https://www.coe.int/en/web/portal/28-january-data-protection-day-factsheet?desktop=true. Accessed 30 Nov 2016.

(b) to obtain at reasonable intervals and without excessive delay or expense confirmation of whether personal data relating to him are stored in the automated data file as well as communication to him of such data in an intelligible form;
(c) to obtain, as the case may be, rectification or erasure of such data if these have been processed contrary to the provisions of domestic law giving effect to the basic principles set out in Articles 5 and 6 of the Convention;
(d) to have a remedy if a request for confirmation or, as the case may be, communication, rectification or erasure as referred to in paragraphs b and c of article 8 is not complied with.

However, this protection is not unrestricted. Indeed, the Convention includes Article 9, which describes what would be the exceptions and restrictions to this common framework:

1. No exception to the provisions of Articles 5, 6 and 8 of this convention shall be allowed except within the limits defined in this article.
2. Derogation from the provisions of Articles 5, 6 and 8 of this convention shall be allowed when such derogation is provided for by the law of the Party and constitutes a necessary measure in a democratic society in the interests of:

   (a) protecting State security, public safety, the monetary interests of the State or the suppression of criminal offences;
   (b) protecting the data subject or the rights and freedoms of others.

3. Restrictions on the exercise of the rights specified in Article 8, paragraphs b, c and d, may be provided by law with respect to automated personal data files used for statistics or for scientific research purposes when there is obviously no risk of an infringement of the privacy of the data subjects.

Thus, Article 9 of the Convention (Exceptions and restrictions) is the most important part of the whole document in terms of CBRNE major crisis situations, as it seems to be perfectly applicable to them. According to this disposition, no exception to the provisions of Articles 5, 6 and 8 of the Convention shall be allowed except within the limits defined in this article. Derogation from the provisions of Articles 5, 6 and 8 of the convention shall be allowed when such derogation is provided for by the law of the Party and constitutes a necessary measure in a democratic society in the interests of protecting State security, public safety, the monetary interests of the State or the suppression of criminal offences, or protecting the data subject or the rights and freedoms of others. Restrictions on the exercise of the rights specified in Article 8, paragraphs b, c and d, may be provided by law with respect to automated personal data files used for statistics or for scientific research purposes when there is obviously no risk of an infringement of the privacy of the data subjects.

In order to make an adequate interpretation of Article 9, the Explanatory Report gives us some important references. At this point, it is stated that exceptions to the basic principles of data protection are limited to those which are necessary for the protection of fundamental values in a democratic society. The text of the second paragraph of this article has been modelled after that of the second paragraphs of

Articles 6, 8, 10 and 11 of the European Human Rights Convention. It is clear from the decisions of the Commission and the Court of Human Rights relating to the concept of "necessary measures" that the criteria for this concept cannot be laid down for all countries and all times, but should be considered in the light of the given situation in each country.

Article 9, paragraph 2.a of the Convention, lists the major interests of the State which may require exceptions (State security, public safety, the monetary interests of the State, or the suppression of criminal offences). The concepts are particularly relevant in a CBRNE situation. These exceptions are very specific to avoid situations where in the general application of the convention, States would have an unduly wide leeway. States retain, under Article 16, the possibility to refuse the application of the convention in individual cases for important reasons, which include those enumerated in Article 9. The notion of *State security*[5] should be understood in the traditional sense of protecting national sovereignty against internal or external threats, including the protection of the international relations of the State.

Thus, one may arrive at a unique conclusion: according to the Convention, data protection and privacy rights could (and probably should) be limited in circumstances when *protecting State security, public safety, the monetary interests of the State or the suppression of criminal offences* makes it necessary. This clause is especially relevant in term of responses against CBRNE major crisis situations, as far as this type of crisis usually involves a great challenge against the goods explicitly included in article 9. Therefore, we could conclude that this Convention allows the signing parties to impose restrictions on data protection and privacy rights when further interests are at stake, even if subject to some substantive limits that should be kept in mind when facing these situations. If we keep in mind that this Convention has been signed and ratified by all EU Member States,[6] we could figure out how important this document could be in the case of addressing a CBRNE major crisis situation.[7]

---

[5] In order to connect the concept of National Security with the European case-law, see European Court of Human Rights-Research Division. 2013. National Security and European case-law. https://rm.coe.int/168067d214. Accessed 30 Nov 2016 (specific references to the Convention for the Protection of Individuals with regard to Automatic Processing of Personal Data of 1981 in pages 37 ff.).

[6] See http://www.coe.int/en/web/conventions/full-list/-/conventions/treaty/108/signatures. Accessed 30 Nov 2016.

[7] Also relevant in this field in the context of the Council of Europe normative production is the Additional Protocol to the Convention for the Protection of Individuals with regard to Automatic Processing of Personal Data regarding supervisory authorities and transborder data flows (Strasbourg, 8-11-2001), particularly its Article 2 (*Transborder flows of personal data to a recipient which is not subject to the jurisdiction of a Party to the Convention*).

12.3 The EU Legal Framework

12.3.1 Data Protection and Privacy as a Fundamental Right in the Charter of Fundamental Rights of the European Union

As it is commonly known, the Charter of Fundamental Rights of the European Union is one of the most important documents ever produced by the EU Institutions. Moreover, it became a legal binding tool when the Treaty of Lisbon[8] was approved in 2007 (entry into force in December 2009). The data protection and privacy issue is addressed in its article number 8:

1. *Everyone has the right to the protection of personal data concerning him or her.*
2. *Such data must be processed fairly for specified purposes and on the basis of the consent of the person concerned or some other legitimate basis laid down by law. Everyone has the right of access to data which has been collected concerning him or her, and the right to have it rectified.*
3. *Compliance with these rules shall be subject to control by an independent authority.*

According to this article, the EU established the right to the protection of personal data as a new fundamental right, distinct from the right to respect for private and family life, home and communications set out in Article 7 of the Charter. This recognised the importance of this issue, which already had been addressed at the time of the redaction of the Charter by the Directive 95/46/EC of the European Parliament and of the Council of 24 October 1995 on the protection of individuals with regard to the processing of personal data and on the free movement of such data.

Article 8 was based on several former legal tools. Firstly, it could be connected with Article 8 of the European Convention on Human Rights (ECHR), even if that document does not explicitly refer to data protection and privacy issues. Its linkage to the Convention for the Protection of Individuals with regard to Automatic Processing of Personal Data (1981), done by the Council of Europe, is also remarkable. It is important to mention that its Preamble included an explicit reference to this issue: "*Considering that it is desirable to extend the safeguards for everyone's rights and fundamental freedoms, and in particular the right to the respect for privacy, taking account of the increasing flow across frontiers of personal data undergoing automatic processing*". More important, its main body constituted as such a big step in the recognition of the right to privacy and data protection as a fundamental one.

Other key documents inspiring the Charter of Fundamental Rights are: the Convention on Human Rights and Biomedicine (1997), especially its Article 10 on 'Private life and right to information'; Article 17 of the International Covenant on Civil and Political Rights (ICCPR, 1966) and the General Comment No. 16 on

---

[8] *Treaty of Lisbon amending the Treaty on the European Union and the Treaty establishing the European Community, signed at Lisbon, 13 December 2007.*

Article 17 ICCPR (especially its paragraph 10 on personal data); Article 12 of the Universal Declaration of Human Rights (UDHR), and the Guidelines for the Regulation of Computerized Personal Data Files adopted by a resolution of the General Assembly of the United Nations on 14th December 1990; and the Additional Protocol to the Convention for the Protection of Individuals with regard to Automatic Processing of Personal Data, regarding Supervisory Authorities and Transborder Dataflow (2001).[9]

The importance of the Charter is that it settles the principles which should guide the construction of both the EU and the Member States' legal framework on data protection and privacy issues. Thus, it must be considered as the principal source of the right to privacy in the EU context and its content must be strictly followed by all legislation addressing this topic. Moreover, EU institutions must make sure that EU citizens' right to gain access to their data and modify them is guaranteed. Currently, this obligation is being accomplished by the European Data Protection Supervisor, *"an independent supervisory authority devoted to protecting personal data and privacy and promoting good practice in the EU institutions and bodies"*.[10] Furthermore, the Article 29 Data Protection Working Party (which was replaced by the European Data Protection Board (EDPB) on the 25 May 2018) was created in order to provide, for instance, expert opinions from Member State level to the Commission on questions of data protection.[11]

In reference to data protection and privacy in CBRNE major crisis situations, article 8.2 of the Charter acquires an extraordinary relevance, as far as its final part opens the gate to the possibility of obtaining data from people without their explicit consent: *"such (personal) data must be processed fairly for specified purposes and on the basis of the consent of the person concerned or some other legitimate basis laid down by law"*. Keeping this part in mind, it seems obvious that the possibility to obtain personal data without the consent of people involved has been recognised by the Charter, under some relevant conditions.[12] The most important is that the exception to the consent must have been included in a law. As we will show later on, it might happen that the restriction is not based on an EU law but on a Member State

[9] EU Network of Independent Experts on Fundamental Rights. June 2006. Commentary of the Charter of Fundamental Rights of the European Union: 91. http://ec.europa.eu/justice/fundamental-rights/files/networkcommentaryfinal_en.pdf. Accessed 30 Nov 2016.

[10] European Data Protection Supervisor. Home. https://secure.edps.europa.eu/EDPSWEB/edps/EDPS?lang=en. Accessed 30 Nov 2016.

[11] European Commission. Justice and Consumers. 2016. Article 29 Working Party. http://ec.europa.eu/justice/data-protection/article-29/index_en.htm. Accessed 30 Nov 2016.

[12] As the EU Network of Independent Experts on Fundamental Rights states, "*The processing of personal data must also be adequate, relevant and not excessive in relation to the purposes for which they were collected and/or further processed. Finally, one of the requirements is that personal data must also be accurate and, where necessary, kept up to date. To guarantee the observance of all these requirements, individuals enjoy legally enforceable rights, notably the right to access and rectify personal data relating to them*" (EU Network of Independent Experts on Fundamental Rights. June 2006. Commentary of the Charter of Fundamental Rights of the European Union: 92–3).

one. This is not relevant in the sense of the Charter: the important matter is that this restriction is established by the applicable legislation and this legal framework describes in sufficient detail appropriate data protection conditions and requirements.

We can, therefore, arrive at an important conclusion: even if data protection and privacy are considered by the Charter as fundamental rights, this does not mean that they could not be limited when they clash with another relevant right or goods, such as, for instance, public health, security, etc. As the Commentary of the Charter of Fundamental Rights of the European Union states, thus providing the reactions of several human rights experts, "*while the EU institutions are under obligation to refuse totally or partly access to a document where disclosure would undermine the protection of personal data, this exception should not be interpreted so as not to disproportionately limit the right of access to documents. Hence, the crucial issue is to define the weight of the respective rights and to optimise the application of each of them, and even so that the right which loses in the process of weighing remain as far as possible relevant in the concrete case*".[13]

Thus, the problem lies in defining the concrete limitations that the statement included in Article 8 about fairness (*data must be processed fairly*) allows. In our opinion, this issue could be redefined as a question over how to balance the defence of privacy and the protection of data with the protection of other relevant values, such as, for instance, security. This is precisely the role to be assumed by legislation developing the Charter, which distinguishes between the different type of data and the circumstances that might let the authorities gain access to them, circumstances which include, as might be guessed, CBRNE major crisis situations.

12.3.2 Processing of Personal Data for the Purposes of the Prevention, Investigation, Detection or Prosecution of Criminal Offences, Including the Safeguarding Against and the Prevention of Threats to Public Security

The changes produced since the approval of Directive 95/46/EC made it necessary to develop an updated version that, gathering the spirit and principles included in that Directive, might face the current legal challenges related to data protection and privacy issues in a much better way. These aims have been addressed through two complementary legislative documents: Regulation 2016/679 of the European Parliament and of the Council on the protection of natural persons with regard to the processing of personal data and on the free movement of such data, and repealing

[13] See EU Network of Independent Experts on Fundamental Rights. June 2006. Commentary of the Charter of Fundamental Rights of the European Union: 92. http://ec.europa.eu/justice/fundamental-rights/files/networkcommentaryfinal_en.pdf. Accessed 30 Nov 2016.

Directive 95/46/EC (General Data Protection Regulation) applicable from 25 May 2018; and Directive (EU) 2016/680 of the European Parliament and of the Council of 27 April 2016 on the protection of individuals with regard to the processing of personal data by competent authorities for the purposes of prevention, investigation, detection or prosecution of criminal offences or the execution of criminal penalties, and the free movement of such data, and repealing Council Framework Decision 2008/977/JHA.

At this level, particular relevance has to be afforded to Directive (EU) 2016/680 of the European Parliament and of the Council of 27 April 2016 on the protection of natural persons with regard to the processing of personal data by competent authorities for the purposes of the prevention, investigation, detection or prosecution of criminal offences or the execution of criminal penalties, and on the free movement of such data, and repealing Council Framework Decision 2008/977/JHA. As its Recital 5 states, Directive 95/46/EC does not apply to the processing of personal data in the course of an activity which falls outside the scope of Community law, such as activities in the areas of judicial cooperation in criminal matters and police cooperation. Consequently, it was considered appropriate for those fields to be addressed by a directive that lays down the specific rules relating to the protection of natural persons with regard to the processing of personal data by competent authorities for the purposes of the prevention, investigation, detection or prosecution of criminal offences or the execution of criminal penalties, including the safeguarding against and the prevention of threats to public security (subject-matter of the Directive according to Article 1.1 Directive 2016/680), respecting the specific nature of those activities. The activities carried out by the police or other law-enforcement authorities include, among others, actions to safeguard against and prevent threats to public security and to fundamental interests of the society protected by law which may lead to a criminal offence.

Those tasks are developed by competent authorities that include not only judicial authorities, the police or other law-enforcement authorities but also any other body or entity entrusted by Member State law to exercise public authority and public powers for the aforementioned purposes. Member States may entrust competent authorities with other tasks which are not necessarily carried out for the purposes of the prevention, investigation, detection or prosecution of criminal offences, including the safeguarding against and the prevention of threats to public security, so that the processing of personal data for those other purposes, in so far as it is within the scope of Union law, falls within the scope of Regulation (EU) 2016/679. As far as CBRNE crises fall in some cases within the scope of the Directive 2016/680, its statements have to be considered where this Directive is relevant. In any case, since (according to Article 2.3 of Directive 2016/680) this Directive should not apply to the processing of personal data in the course of an activity which falls outside the scope of Union law, activities concerning national security, activities of agencies or units dealing with national security issues and the processing of personal data by the Member States when carrying out activities which fall within the scope of Chapter 2 of Title V of the Treaty on European Union (TEU) should not be considered to be activities falling within the scope of this Directive.

In order to ensure the same level of protection for natural persons through legally enforceable rights throughout the Union and to prevent divergences hampering the exchange of personal data between competent authorities, Directive 2016/680 should provide for harmonised rules for the protection and the free movement of personal data processed for the purposes of the prevention, investigation, detection or prosecution of criminal offences or the execution of criminal penalties, including the safeguarding against and the prevention of threats to public security. The approximation of Member States' laws should not result in any lessening of the personal data protection they afford but should, on the contrary, seek to ensure a high level of protection within the Union. The Member States should not be precluded from providing higher safeguards than those established in this Directive for the protection of the rights and freedoms of the data subject with regard to the processing of personal data by competent authorities.

As CBRNE situations can give rise to the need to collect or gain access to personal data concerning health or to genetic data it is important to point out that article 3 of Directive 2016/680 provides a definition of genetic data (Article 3.12) and of personal data concerning health (Article 3.14). According to this Directive, personal data should be collected for specified, explicit and legitimate purposes within the scope of the Directive and should not be processed for purposes incompatible with the purposes of the prevention, investigation, detection or prosecution of criminal offences or the execution of criminal penalties, including the safeguarding against and the prevention of threats to public security. The performance of the tasks of preventing, investigating, detecting or prosecuting criminal offences or preventing threats to public security allows the competent authorities to require or order natural persons to comply with requests made in order to obtain personal data. In those cases the consent of the data subject as defined in Regulation 2016/679 should not provide a legal ground for processing personal data by competent authorities as where the data subject is required to comply with a legal obligation, the data subject has no genuine and free choice.

Article 4 of Directive 2016/680 sets out the principles relating to the processing of personal data. Particularly interesting is Article 4.2 of Directive 2016/680 which states that "processing by the same or another controller for any of the purposes set out in Article 1.1 other than that for which the personal data are collected shall be permitted in so far as: (a) the controller is authorised to process such personal data for such a purpose in accordance with Union or Member State law; and (b) processing is necessary and proportionate to that other purpose in accordance with Union or Member State law".

Related to specific processing conditions Article 9 of Directive 2016/680 states that personal data collected by competent authorities for the purposes set out in Article 1(1) shall not be processed for purposes other than those set out in Article 1(1) unless such processing is authorised by Union or Member State law. Where personal data are processed for such other purposes, Regulation (EU) 2016/679 shall apply unless the processing is carried out in an activity which falls outside the scope of Union law.

Connected with the information to be made available or to be given to the data subject, particularly relevant are the restrictions set out in Article 13.3 of Directive 2016/680 which states that Member States may adopt legislative measures delaying, restricting or omitting the provision of the information to the data subject pursuant to Article 13 paragraph 2 to the extent that, and for as long as, such a measure constitutes a necessary and proportionate measure in a democratic society with due regard for the fundamental rights and the legitimate interests of the natural person concerned, in order to: (a) avoid obstructing official or legal inquiries, investigations or procedures; (b) avoid prejudicing the prevention, detection, investigation or prosecution of criminal offences or the execution of criminal penalties; (c) protect public security; (d) protect national security; or (e) protect the rights and freedoms of others. Important limitations to the right of access to the data subject are also established on similar grounds in Article 15 Directive 2016/680.[14]

Directive 2016/680 also stipulates the conditions under which a transfer of personal data to a third country or to an international organisation connected with the prevention, investigation, detection or prosecution of criminal offences or the execution of criminal penalties, including the safeguarding against and the prevention of threats to public security, can take place.

12.3.3 Regulation (EU) 2016/679 of the European Parliament and of the Council of 27 April 2016 on the Protection of Natural Persons with Regard to the Processing of Personal Data and on the Free Movement of Such Data, and Repealing Directive 95/46/EC (General Data Protection Regulation)

Regulation 2016/679 includes three main innovations:

1. One continent, one law: The Regulation will establish a single, pan-European law for data protection, replacing the current inconsistent patchwork of national laws. Thus, companies operating in the EU territory will deal with one law, not 28.
2. One-stop-shop: The Regulation will establish a 'one-stop-shop' for businesses so companies will only have to deal with one single supervisory authority, not 28, making it simpler and cheaper for companies to do business in the EU.
3. The same rules for all companies – regardless of their establishment: today European companies have to adhere to stricter standards than their competitors established outside the EU but also doing business on the EU Single Market.

[14] See also Article 16.4, Directive 2016/680 which on the basis of the same grounds gives the possibility for the Member States to adopt legislative measures restricting, wholly or partly, the obligation to provide for the controller to inform the data subject in writing of any refusal of rectification or erasure of personal data or restriction of processing and of the reasons of the refusal.

> With the reform, companies based outside of Europe will have to apply the same rules. European regulators will be equipped with strong powers to enforce this: data protection authorities will be able to fine companies who do not comply with EU rules with up to 2% of their global annual turnover. European companies with strong procedures for protecting personal data will have a competitive advantage on a global scale at a time when the issue is becoming increasingly sensitive.[15]

Due to the fact that the protection of natural persons with regard to the processing of personal data by competent authorities for the purposes of the prevention, investigation, detection or prosecution of criminal offences or the execution of criminal penalties, including the safeguarding against and the prevention of threats to public security and the free movement of such data, is the subject of a specific Union legal act (namely Directive 2016/680), Regulation 2016/679 should not, therefore, apply to processing activities for those purposes. Nevertheless, Member States may entrust competent authorities within the meaning of Directive (EU) 2016/680 with tasks which are not necessarily carried out for the purposes of the prevention, investigation, detection or prosecution of criminal offences or the execution of criminal penalties, including the safeguarding against and prevention of threats to public security, so that the processing of personal data for those other purposes, in so far as it is within the scope of Union law, falls within the scope of Regulation 2016/679.

With regard to the processing of personal data by those competent authorities for purposes falling within the scope of this Regulation, Member States should be able to maintain or introduce more specific provisions to adapt the application of the rules of this Regulation. Such provisions may determine more precisely specific requirements for the processing of personal data by those competent authorities for those other purposes, taking into account the constitutional, organisational and administrative structure of the respective Member State. When the processing of personal data by private bodies falls within the scope of Regulation 2016/679, this Regulation should provide for the possibility for Member States under specific conditions to restrict by law certain obligations and rights when such a restriction constitutes a necessary and proportionate measure in a democratic society to safeguard specific important interests including public security and the prevention, investigation, detection or prosecution of criminal offences or the execution of criminal penalties, including the safeguarding against and the prevention of threats to public security.

According to Recital 73 Regulation 2016/679 restrictions concerning specific principles and the rights of information, access to and rectification or erasure of personal data, the right to data portability, the right to object, decisions based on profiling, as well as the communication of a personal data breach to a data subject

[15] European Commission Memo, 2014. Progress on EU data protection reform now irreversible following European Parliament vote. *European Commission Press Release Database*, March 12. http://europa.eu/rapid/press-release_MEMO-14-186_en.htm. Accessed 30 Nov 2016.

and certain related obligations of the controllers may be imposed by Union or Member State law, as far as necessary and proportionate in a democratic society to safeguard public security, including the protection of human life especially in response to natural or manmade disasters, the prevention, investigation and prosecution of criminal offences or the execution of criminal penalties, including the safeguarding against and the prevention of threats to public security, or of breaches of ethics for regulated professions, other important objectives of general public interest of the Union or of a Member State, in particular an important economic or financial interest of the Union or of a Member State, the keeping of public registers kept for reasons of general public interest, further processing of archived personal data to provide specific information related to the political behaviour under former totalitarian state regimes or the protection of the data subject or the rights and freedoms of others, including social protection, public health and humanitarian purposes (see Article 23 – Restrictions). Those restrictions should be in accordance with the requirements set out in the Charter and in the European Convention for the Protection of Human Rights and Fundamental Freedoms.

12.4 Conclusion

The aim of the Chapter was to give a general overview of the current legal framework on Data Protection in CBRNE crises, both at Council of Europe and EU level. At this point it is relevant to highlight that even when CBRNE crisis situations are not a major issue when regulating data protection, some important conclusions can be drawn for the processing of personal data in the context of CBRNE crises by means of the exceptions established in the general regulation on personal data protection issued by the Council of Europe (Convention 108) and the EU (Charter of Fundamental Rights of the European Union, Directive 2016/680 and Regulation 2016/679). Particularly, and on the basis of the protection of, among others, public security and public safety, it is possible in the context of the legal framework mentioned above, to establish restrictions to the general criteria governing personal data protection and privacy rights.

References

Council of Europe. The Council of Europe in brief. http://www.coe.int/en/web/about-us/who-we-are. Accessed 30 Nov 2016.

Council of Europe. Modernisation of the data protection "Convention 108". https://www.coe.int/en/web/portal/28-january-data-protection-day-factsheet?desktop=true. Accessed 30 Nov 2016.

EU Network of Independent Experts on Fundamental Rights. June 2006. Commentary of the Charter of Fundamental Rights of the European Union: 91. http://ec.europa.eu/justice/fundamental-rights/files/networkcommentaryfinal_en.pdf. Accessed 30 Nov 2016.

European Commission Memo. 2014. Progress on EU data protection reform now irreversible following European Parliament vote. *European commission press release database*, March 12. http://europa.eu/rapid/press-release_MEMO-14-186_en.htm. Accessed 30 Nov 2016.
European Commission. Justice and Consumers. 2016. Article 29 Working Party. http://ec.europa.eu/justice/data-protection/article-29/index_en.htm. Accessed 30 Nov 2016.
European Court of Human Rights-Research Division. 2013. National security and European case-law. https://rm.coe.int/168067d214. Accessed 30 Nov 2016.
European Data Protection Supervisor. Home. https://secure.edps.europa.eu/EDPSWEB/edps/EDPS?lang=en. Accessed 30 Nov 2016.

Chapter 13
Directive 95/46/EC and Protection of Personal Data in CBRNE Events: Limits and Perspectives

Anna Falcone

Abstract Directive 95/46/EC of the European Parliament and of the Council of 24 October 1995 on the protection of individuals with regard to the processing of personal data and on the free movement of such data was a milestone in the regulation of personal data protection in the European Union (EU). Despite subsequent legislation, it remains a major step forward which must be examined in any comparative and evolutionary analysis of EU legislation on personal data processing. This directive established, in fact, the core framework for personal data protection and the rights of the concerned parties, while enforcing independent supervision by national authorities. This chapter is aimed at identifying its fundamental ideas and the main issues involved.

Keywords Data protection regulation · Data privacy · Informed consent and data sharing · Circulation of data · Data processing

13.1 The Protection of Personal Data in the Legal Framework of the EU

The protection of personal data and the right to privacy, and private and family life are fundamental rights, provided for in the legal framework of the European Union (EU) and of its Member States. Said rights were explicitly afforded in several programmatic and statutory documents, in particular in Articles 7 and 8 of the Charter of Fundamental Rights of the European Union and Article 16 of the Treaty on the Functioning of the European Union (TFEU), which now constitute the legal basis for future policies and legislation within the EU.

The Lisbon Treaty provides a stronger basis for the development of a clear and efficient personal data protection system, while giving new powers to the European Parliament. The Charter of Rights raises the level of protection of personal data to

A. Falcone (✉)
Studio Legale Falcone, Rome, Italy

D. P. O'Mathúna, I. de Miguel Beriain (eds.), *Ethics and Law for Chemical, Biological, Radiological, Nuclear & Explosive Crises*, The International Library of Ethics, Law and Technology 20, https://doi.org/10.1007/978-3-030-11977-5_13

the rank of fundamental rights. This represents a considerable jump, compared to earlier regulations, and reinforces the weight given to such rights in balance with the rights of equal rank in EU policies on security, data communications, freedom, crime prevention and international relations. The protection of individual rights is, however, in constant conflict with the demands of the free market, which often considers policy as a 'commodity,' and with public safety needs, especially in the prevention of and reaction to chemical, biological, radiological, nuclear and explosive (CBRNE) events. Moreover, in a global society characterized by rapid technological changes, where information exchange knows no borders, regulation of this field is destined to age rapidly. Therefore, an effective regulatory framework requires constant updates. The escalation of the threats faced by modern society and the development of increasingly sophisticated technological tools to collect, process, and transmit personal data, and place them into public and private databases, makes it particularly difficult to find an acceptable balance between strengthening security and protecting human rights, including the protection of personal data and privacy.

At present, the most important relevant acts in the European Union regulatory framework are: Directive 95/46/EC on the protection of individuals with regard to the processing of personal data and on the free movement of such data,[1] the Directive on e-privacy and electronic communications (amended in 2009),[2] Directive 2006/24/EC on data retention (declared invalid by the Court of Justice of the European Union on 8 April 2014),[3] Regulation (EC) No 45/2001 on the protection of individuals with regard to the processing of personal data,[4] and the Council Framework Decision 2008/977/JHA of 27 November 2008 on the protection of personal data processed in the framework of police and judicial cooperation in criminal matters.[5] At first sight, this might look like a quite impressive framework. However, experience shows that in recent years this framework has been overtaken by rapid technological developments and needs urgent improvement to provide greater protection for the rights of the people. This is, in fact, an ongoing process involving an overall review of EU legislation. Nevertheless, old rules will remain in force until their substitution with new legislation, mainly the Regulation of the European Parliament and of the Council on the protection of individuals with regard to the processing of personal data and on the free movement of such data (General Data Protection Regulation). Such initiatives are explored in Chap. 12 which describes the current legal framework on data protection in CBRNE crises.

[1] O.J. (L 281) 23.11.1995, 31.

[2] O.J. (L 201), 31.7.2002, 37.

[3] O.J. (L 105) 13.4.2006, 54.

[4] O.J. (L 008) 12/01/2001, 1.

[5] O.J. (L 350) 30/12/2008, 60.

13.2 Directive 95/46/EC and the Regulation on Data Protection

Directive 95/46/EC of the European Parliament and of the Council of 24 October 1995 on the protection of individuals with regard to the processing of personal data and on the free movement of such data was a milestone in the regulation of personal data protection in the EU. Despite subsequent legislation, it remains a major step forward which must be examined in any comparative and evolutionary analysis of EU legislation on personal data processing. The legislation relevant to CBRNE which was enacted after this directive is examined in Chap. 12 which follows this one. This directive established, in fact, the core framework for personal data protection and the rights of the concerned parties, while enforcing independent supervision by national authorities. It is a very flexible regulation, in the formal aspects of its substance, namely the implementation and interpretation of its regulatory content.

The Directive aims to guarantee the protection of the rights and fundamental liberties of the individual, and particularly the right to privacy regarding the processing of personal data. Article 2 identifies and defines the legal meaning of "personal data", "processing of personal data", "personal data filing system", "controller", "processor", "third party", "recipient", and "the data subject's consent", a set of concepts which still constitutes the most relevant legal categories in data protection regulation. Article 3, however, restricts the scope of this Directive to "the processing of personal data wholly or partly by automatic means, and to the processing otherwise than by automatic means of personal data which form part of a filing system or are intended to form part of a filing system." Instead, cases relating to activities which fall outside the scope and competence of EU law, such as those related to public safety, defence, national security, and activities of the Member States in criminal matters, are specifically excluded from the scope of the Directive. In this context, therefore, the choices related to critical situations and exceptional events will be taken by the Council and will not be subject, for express restriction of the Treaties, to the Union's common rules.

Beyond these cases, the Directive leaves ample space for national legislation which remains prevalent when the processing of personal data is carried out in the territory of one or more Member States, or in a territory subject to national sovereignty (Article 4). To this must be added that, according to Article 5, Member States are empowered to determine the conditions under which the processing of personal data is lawful, even if within the limits of the provisions made by the Directive. Indeed, the Directive merely sets down general rules for the treatment, detection and update of personal data and the objective and subjective conditions of legitimacy under the laws of the Union, by delegating to the State to implement and take effect these conditions (Article 6). It is inevitable, therefore, that each Member State adopts different approaches to accomplish these objectives, whereby it is difficult – if not impossible – to ensure uniform fulfilment of these procedures, or to conduct a centralized analysis of these data, when they are relevant or become such, to prevent or mitigate dangerous situations.

Even Article 7 – which sets down the principles of "legitimacy of data processing" – is open to different interpretations, a circumstance that led to huge variations in its specific implementation at a national level. The general rule of "explicit and unequivocal consent" of the person concerned, as a primary condition legitimizing the processing of personal data, has severe problems. Consequently, it is even possible to wonder whether its protection is not just a smokescreen to justify, on the contrary, the many "exceptions to the rule" endorsed by Directive. Indeed, in the case of major crisis situations, relevant exceptions to the general legal framework can be found in par. 1, lett. (d) and (e), which justify data processing *even in the absence of consent,* provided that it is necessary to protect the vital interests of the concerned person, or to accomplish a public duty inherent to the performance of a public function, or to perform a task in the public interest or in the exercise of public function that corresponds to the controller (a third party to which the data are disclosed). The exemption from individual consent established by this clause seems fully justified with regards to the preservation of a vital interest of the individuals. However, the limits on the exception may be interpreted in diverse ways by national laws, a circumstance that fuels the risk of sacrificing fundamental rights or primary personal interests for reasons difficult to verify. One must consider, in this regard, the considerably diverse meanings that concepts such as "biosecurity" or "public interest" might acquire from Member State to Member State, and the different public programs or measures that might be put into practice to monitor, prevent, or curb situations of real or presumed danger.

One of the most remarkable initiatives in the Directive is the ban on processing of particular data categories classified as "sensitive data" (Article 8). These are data concerning health or sexual life, data that can reveal racial or ethnic origin, political opinions, religious or philosophical beliefs, or trade union membership. However, there are some exceptions to this protection. In general, the prohibition is not applicable if the person gives his/her explicit consent to the processing of data, or if he/she has otherwise made them public. There is, however, an exception particularly interesting in the case of a CBRNE incident. The Directive explicitly states in par. One letter. (c) that protection of sensitive data will not be applicable "if the processing is necessary to protect the vital interests of the person concerned or of a third party in the event that the subject is physically or legally incapable of giving consent".

The Directive provides (Article 8.4) another exception to data protection for the prevailing interests of the Member States in high-risk situations. In these cases, the Member States are authorized to allow processing of sensitive personal data, even in the absence of an explicit consent from the concerned person. According to the Recital n. 34, the privileged sectors in which this exception should operate are "Public Health and Social Protection." The objectives which justify this exception are to "ensure the quality and cost-effectiveness of procedures used for settling claims for benefits and services in the health insurance system – scientific research and government statistics." In such cases, it is considered an essential responsibility of the States to provide the appropriate specific safeguards to protect the fundamental rights and the privacy of their citizens. This obligation is not always upheld, so that personal rights are eventually protected in a relative and pliable way. This is particularly relevant since this exception can be implemented either by a legislative measure – that is, with the guarantees of the legal reserve provided for the protection of fundamental rights – or by a decision of the supervisory authority, which does not provide the same guarantees. Therefore, the Directive permits a clear discrepancy in the levels of privacy protection guaranteed to the citizens of different Member States, depending on the choices made by their governments.

Nevertheless, and to be fair to the Directive, one must concede that it attempts to ensure a uniform protection of these rights and to make actionable the right to privacy and control of personal data. Its Articles 10 and 11 list in detail the information that has to be provided to the person concerned if his/her personal data is collected, no matter whether collected directly or through third parties. It states that people involved "must have access to information relating to: a) the identity of the controller, purpose of the data processing, the recipients or categories of recipients, voluntary or otherwise of responses and possible consequences, and the right to access and correct the data". The only exceptions – such as collection for statistical purposes, or for historical or scientific research – are those in which this access could not possibly be provided or would involve a disproportionate effort, apart from the cases in which recording or disclosure was required by law. Under those circumstances, Member States are obliged to provide appropriate safeguards. Furthermore, Article 12 guarantees every concerned person his/her right to access his/her own data, which are to be allowed in a free, intelligible, and full way without restrictions, so that it might be possible to know its origin and request its modification, deletion or blocking (also for third parties).

13.3 Article 13 and the Specific Provisions on Major Crises

The previous points can be summarized in one simple idea: even if the Directive was intended to protect individual privacy, this protection could be limited in exceptional circumstances such as those present in major crises. However, the most significant provision related to emergency or exceptional events, is contained in Article 13. It states that, in cases of emergency, the Member States may adopt legislative measures to restrict the scope of the obligations and rights provided for in Articles 6 (1), 10, 11 (1), 12 and 21 of the Directive, "when such a restriction constitutes a necessary measures to safeguard: (a) national security; (b) defense; (c) public security; (d) the prevention, investigation, detection and prosecution of criminal offences, or of breaches of ethics for regulated professions; (e) an important economic or financial interest of a Member State or of the European Union, including monetary, budgetary and taxation matters; (f) a monitoring, inspection or regulation function connected, even occasionally, with the exercise of official authority in cases referred to in (c), (d) and (e); (g) the protection of the data subject or of the rights and freedoms of others."

Therefore, this concrete clause questions the entire regulatory framework of the Directive regarding the rights protected, putting in the hands of the Member States the effectiveness of the right to privacy and the effective control over personal data. On this basis, Member States are, in fact, the only entities entitled to evaluate the occurrence of actual or perceived danger conditions which might trigger the so-called "exceptional regime", that legitimizes the compression of those rights and the corresponding warranty obligations, especially in cases of safeguarding "State security", "defence", or "public security". As a consequence, Member States exercise an almost absolute power, strengthened by the unquestionableness of the policy choices made by governments in cases of, for example, terrorist attacks, natural disasters or exceptional events that threaten the lives of people. The only legal limits enforceable in such cases are those embedded in the principles of proportionality and respect for fundamental rights, as codified by national Constitutions, EU law, the European Convention on Human Rights and International Treaties.

The unavoidable consequence of the existence of so many "exceptional regimes" in different Member States in emergency situations – and particularly during CRBNE crises – is clear: no common EU legal framework ensures an adequate conciliation between the rights of citizens and the third parties involved. Thus, it must be concluded that the coexistence of different national regulations and different interpretations resulted in an unsatisfactory situations for all parties involved.

The situation changed after the entry into force of the Lisbon Treaty and the recognition of a common citizenship, which needed to be reflected in the efficient guarantee of the rights established in the Charter and in the EU treaties. It then became necessary and indefectible to introduce a common binding regulation to assure all European citizens a univocal level of protection for privacy and clear limits to the availability of their personal data to the Member State or third parties, especially in those cases where the rights enforced by the Charter and the EU treaties were functional or relating to the protection of other fundamental rights. This is precisely the scenario that arises in crisis situations and exceptional circumstances, such as CBRNE events, when protection needs emerge with urgency. Of course, this protection cannot be limited to the EU sovereign territory. Indeed, it would not be realistic at all to think that in our globalized world, where personal data runs on the international "information highway" and are commonly handled by multinational companies or other private and public entities outside the EU, that one can continue to adhere to the territoriality principle.

It was precisely with international movement of data in mind that Article 25 of the Directive establishes, as a general rule, that the transfer of personal data to third countries can only take place if the Country in question afforded "an adequate level of protection", giving particular attention to the nature of the data, the purpose and duration of the proposed processing operation or operations, the Country of origin and Country of final destination, the rules of law (both general and sectorial) in force in the third Country in question and the professional rules and security measures which exist in that Country. However, the determination of the "adequate level of protection" is left – also in this case – to the Member States, even if the norm establishes a form of agreement between the Member States and the Commission, which should gather and act together when, in their opinion, a third Country does not guarantee an adequate level of protection. However, this proceeding situation occurs only *ex post*, so it is unable to prevent the public disclosure of the information provided by a Member State, based on a undetailed evaluation, which could result in a leak of information and data that would be very difficult to recover. The remedy provided by par. 5 – the opening of negotiations by the Commission with the third country in question – is clearly unhelpful, due to its delayed and inadequate application, especially in the case of exceptional events or crises requiring quick decisions and immediate protection measures.

The general rule in Article 25 is also subject to numerous exceptions which, once again, feeds the doubt – if not the certainty – that Directive 95/46 was conceived more to legitimize the exceptions than to protect rights. The peculiarity of these exceptions is the transfer of the burden of data protection from the public to the private level. In fact, par. 2 allows data transfer to Countries that do not ensure an adequate standard of protection, provided that the controller gives sufficient guarantees for the protection of privacy, rights and fundamental freedoms of the concerned people, as well as for the exercise of the rights involved. The replacement of public institutions by private ones implies a lowering of the general level of the protection of personal rights, precisely in circumstances in which the underlying rights deserve more protection, that is, in emergency and crisis situations, such as CBRNE events. Even the planned opposition procedure against the authorizations granted in derogation are late and ineffective. Both internal security, and the rights of concerned people, could only be effectively protected *ex ante* and before their endangerment or violation. The reinstatement and any *ex post* compensations must be considered, especially in this matter, remissive instruments of the law, which come into play when the safety standard has failed, and neither security nor rights are adequately protected.

13.4 The Jurisprudence of the European Court of Justice

Beyond this 'static' analysis, a critical analysis of the personal data legal framework set by Directive 95/46 easily reveals challenging issues in its 'dynamic' application. It is indeed in judicial rulings where all the limitations of the data protection regulation emerged. We provide a concise list here. To begin with, it makes sense to remember the judgment of the European Court of Justice, Case C-553/07 of 7 May 2009, regarding access to personal data, in relation to Article 12 lett. a), which sanctioned the asymmetry between the duration and the exercise of people's right of access to their own data and the obligation entrusted to the controller to retain them for an extended period of time.

In the same sense, the decision of the Court of Justice (Joined Cases C-293/12 and C-594/12), dated 8 April 2014 – the so-called "*data retention*" ruling – in relation to the safety and protection of data stored by the service providers of electronic communications of public access and the public telecommunication networks, established that Directive 95/46/CE "entails a wide-ranging and particularly serious interference with those fundamental rights in the legal order of the EU, without such an interference being precisely circumscribed by provisions to ensure that it is actually limited to what is strictly necessary (…) in order to ensure their full integrity and confidentiality. Furthermore, a specific obligation on Member States to establish such rules has also not been laid down".[6] Even worse, the Court stated that "*the* directive does not require the data in question to be retained within the European Union, with the result that it cannot be held that the control, explicitly required by Article 8(3) of the Charter, by an independent authority of compliance with the requirements of protection and security, as referred to in the two previous paragraphs, is fully ensured".[7] Based on these and other reasons, Directive 2006/24/EC of the European Parliament and of the Council dated 15 March 2006 on the retention of data generated or processed in connection with the provision of publicly available electronic communications networks or public communications and amending Directive 2002/58/EC, was declared invalid. This left a loophole in the legal framework of the EU, which needs more updated and coherent regulations based on the levels of protection recognized by the Charter and the Treaties.

Lastly, the decision of the Court of Justice, dated 13 May 2014 (the so-called "*Google case*" – C131/12), which extended a case regarding the processing of personal data to the results of the search engines, and provided an "authentic interpretation" of the rights afforded by Directive 95/46/CE, widened the rights of the concerned parties regarding the availability of their data, recognizing a true "*right to be forgotten*". According to the ruling, search engines must guarantee this right, deleting, at the request of the affected person, the results of a search and the links that would take them to the personal data related to his private life, affording the protection of his right to privacy and the fundamental rights connected to it. Among those rights must be highlighted Articles 7 and 8 of the Charter of Fundamental Rights, which, according to the Court, must initially prevail over the economic interest of the operator of the search engine, and over the interest of the public in having access to that information, unless the public profile of the affected person justifies a balance between his or her rights and the public interests at stake.

[6] N. 65 and 66 of the Ruling.

[7] N. 68 of the Ruling.

13.5 Conclusions and Perspectives for a New Regulation of Data Protection

In conclusion, the Directive of the EU of 1995 marked an important step in the history of data protection. Its objectives – "*to ensure the flow of data and the effective protection of the rights and liberties of the individuals*" – are still valuable, but they need to be adapted to current times. In a globalized and digital world that has made "speed" its most distinguishing trait, the present regulation in force does not afford the necessary degree of harmonization between the national statutory laws and the European rules that apply to such scenarios. Moreover, several issues weaken the internal coherence of a regulatory system that really seeks to balance the effective protection of personal data with the demands of security and the needs of the market. First, a complex system of exceptions exists at different levels. Second, the fact that the guarantors for many of the granted rights of individuals are private companies is a circumstance that hinders adequate control of the level of protection effectively granted. Last, but not least, it must be highlighted that the specific problems of processing personal data, especially about bio-security, terrorist attacks and exceptional situations of crisis, are not sufficiently integrated into the EU Action Strategy on CBRN.[8]

The disappearance of the "institutional pillar structure" of the European Union following the introduction of the Lisbon Treaty, aside from providing a stronger foundation for the system of data protection, created the conditions for a more uniform and efficient regulation. Furthermore, the Commission, as well as the Council and the Parliament, identified, as a key goal, better protection of personal data within the European Union. The main results of this new scenario have been realized already. On 25 January 2012, the European Commission introduced a comprehensive new regulation to update the existing legal framework of the EU on data protection. This proposed discussion on: a) a proposal for a General Data Protection Regulation, which replaces the Directive 95/46 setting up a general European framework on data protection; b) a proposal for a directive of the European Parliament and of the Council on the protection of individuals with regard to the processing of personal data by competent authorities for the purposes of prevention, investigation, detection or prosecution of criminal offences or the execution of criminal penalties, and the free movement of such data.

After a long debate, the official texts of the Regulation and the Directive were published in the EU Official Journal on 4 May 2016.[9] The Regulation entered into force on 24 May 2016 and will be applied from 25 May 2018. The Directive came into force on 5 May 2016 and EU Member States have to transpose it into their national law by 6 May 2018. This new legal framework aims to strengthen citizens'

---

[8] "*Communication from the Commission to the European Parliament and the Council of 24 June 2009 on Strengthening Chemical, Biological, Radiological and Nuclear Security in the European Union*" (http://ec.europa.eu/justice_home/news/sumary/docs/com_2009_0273_en.pdf).

[9] Official Journal of the European Union, L 119, 4 May 2016.

control over their personal data and – at the same time – to simplify the regulatory environment for business and improve access to the digital economy. In this way, it will become an important guarantee for the fundamental rights recognized there and an embankment to their violation, even in crisis situations, as Chap. 12 in this volume shows. However, its clauses will have to be integrated with the decisions of the Council on exceptional crisis events and with international agreements on the exchange of personal data and information, especially in the areas of global counter-terrorism and major criminal threats.

References

Official Journal of the European Union, L 281, 23 November 1995.
Official Journal of the European Union, L 201, 31 July 2002.
Official Journal of the European Union, L 105, 13 April 2006.
Official Journal of the European Union, L 008, 12 January 2001.
Official Journal of the European Union, L 350, 30 December 2008.
Official Journal of the European Union, L 119, 4 May 2016.
Court of Justice, Joined Cases C-293/12 and C-594/12, 8 April 2014.
Court of Justice, C131/12, 13 May 2014.
European Commission. Strengthening chemical, biological, radiological and nuclear security in the European Union. 24 June 2009. Communication to the European Parliament and the Council. http://ec.europa.eu/justice_home/news/sumary/docs/com_2009_0273_en.pdf.

Chapter 14
Triage Issues in a CBRNE Crisis: Experiences from European Projects

Catherine Bertrand, Eric Lecarpentier, Francis Herodin, and Frederic Dorandeu

Abstract For any major incident including chemical, biological, radiological, nuclear and explosive (CBRNE) events with casualties, there is a requirement to allocate resources on a priority basis whenever resources are outweighed by demand. This is based on the sorting of casualties so that the greatest good is provided to the greatest number of casualties. Triage is thus vital for ensuring the success of disaster management. Success relies on previous training in applying specific plans.

One key question is to determine if the event results from an accident or from a terrorist action. A multi-site attack would impact the allocation of resources to a given site. Depending on the agent, CBRNE events have different kinetics and consequences for FR (first responder) safety. Triage is part of the medical response to the incident that also requires cordoning, appropriate command and control, communications, assessment and hazard management.

Triage represents not only a technical issue but also involves ethical questions that must be taken into account, before and as soon as the event occurs. Providing real time triage, trustworthy guidance can avoid rumours and wrong behaviours. It is mandatory to rapidly track all casualties. The psychological impact of a CBRNE event should also be taken into consideration in order to minimize post-traumatic stress disorder.

Efforts should be made to raise awareness of the importance of practical triage training and preparation before sending health-care professionals to the scene of a disaster, in particular in the case of CBRNE attacks.

C. Bertrand (✉) · E. Lecarpentier
Service d'Aide Médicale Urgente (SAMU), SAMU 94 CRETEIL, CRETEIL, France
e-mail: catherine.bertrand2@aphp.fr

F. Herodin
French Armed Forces IRBA, Bretigny sur Orge, France

F. Dorandeu
French Armed Forces Biomedical Research Institute IRBA, Bretigny sur Orge, France

D. P. O'Mathúna, I. de Miguel Beriain (eds.), *Ethics and Law for Chemical, Biological, Radiological, Nuclear & Explosive Crises*, The International Library of Ethics, Law and Technology 20, https://doi.org/10.1007/978-3-030-11977-5_14

Keywords Triage · Ethics · CBRNE threats · Healthcare · EDEN project · Mass casualties

14.1 Triage Issues in a Disaster Situation

14.1.1 Triage Serves Safety

For any major incident including chemical, biological, radiological, nuclear and explosive (CBRNE) events the sorting of casualties so that the greatest good is provided to the greatest number of casualties is of upmost importance. Appropriate planning for medical support of such events comprises anticipation and analysis of threats and disaster circumstances with the objective of preserving the overall societal organisation.

From a medical point of view, we, as civilian and military practitioners define a disaster situation as a situation needing extra countermeasures and means compared with a daily emergency situation. For example, our emergency medical service (SAMU), which is one of the French pre hospital medical services and dispatch centres, initiates special plans as soon as extra medical teams are needed to face a situation. Our SAMU has six mobile intensive care units (ICU) with medical teams available on a daily basis for a population of 1.6 million inhabitants. In the case of an unexpected, overwhelming event, SAMU needs to ask for help and for this, applies special contingency pre hospital plans according to the situation. Similarly, if the hospital has to receive many casualties, another special plan called the "white plan" is initiated. The goal of the "white plan" is to split internal medical and logistical resources in order to take care of regular patients and injured people at the same time.

Emergency departments follow a routine triage system in order to cope with overcrowding (Aacharya et al. 2011). In a disaster situation, hospitals are vulnerable and pre hospital triage is aimed to preserve them, in order to let them take care of all casualties without affecting the care of other daily emergencies. For example if you have a heart attack or if you are giving birth, you should not be affected by the management of the ongoing crisis.

Indeed, and on an ethical point of view, dealing with the exceptional situation must not impact the quality of the treatment of regular emergencies. The goal of our dispatch centres (SAMU) is to sort out patients from the site to the available hospitals while preserving nearby hospitals. A table with available beds and a map of nearby hospitals is provided in advance and updated accordingly. This table is used by the medical commander on the site.

On the scene, basic principles of the medical response to a CBRNE incident require cordoning, appropriate command and control, communications, assessment and hazard management. They have been well described in a recent NATO document (AMedP7.1, NATO 2015).

In the face of an unexpected mass event, we have to choose between two approaches: scoop and run or stay and give early care on the site in an advanced medical post. The answer is in between thanks to adapted triage procedures on the spot. If we fail to perform rapid and efficient triage, hospitals will be overcrowded with unstable casualties. These triage principles are the key factors for success and explain why regular patients should not be affected by the crisis situation.

On site triage goals comprise all the listed items below:

- Assess the severe cases at a glance
- Tag and trace all people involved
- Dispatch victims towards the adapted medical units and hospitals
- Protect nearby hospitals from a sudden overload of work

In a CBRNE situation, the first step is to get situational awareness which should provide a global view shared in real time between different headquarters via reliable communication means and secured information networks. This situation should be transmitted to first responders (FRs) and the general population, in order to help them to be aware of what is happening and can be done at the earliest stage, including health recommendations and relevant behaviours. The method of communication and communication networks is the key factor for success. Communications should be reliable, interoperable and fast to deploy. Sharing information in real time between all involved is mandatory.

The second step is related to safety and security issues which should be provided to the population and FRs equipped with suitable personal protective equipment (PPE).

The third step represents safe health for the population, including the management of casualties with fast medical triage and appropriate treatment at scene and in hospitals.

All these procedures aim at providing safety and health care for casualties and rescue service members at the same time. Safety measures are, in fact, the responsibility of politicians and decision-makers and depend on availability of protective equipment, training of rescue services and information of the general public. The goal of the crisis management is to take care of victims and to keep FRs safe while they perform their tasks.

Of course a disaster response depends on the size of the event, the number of people involved and accordingly the number and the capacity of FRs to manage the crisis. In a large city like Paris emergency services have as many FRs as potential casualties. So it might not be ethical to leave these above objectives unmet.

Triage should be flexible and adaptable because we are constantly facing unexpected events, most probably in a hostile environment, dealing with a large number of casualties with unusual pathologies, who are usually difficult to access.

If CBRNE injuries are combined with conventional ones caused by a blast, bullets or a crush, it worsens the situation.

Another situation we should be constantly aware of is the possibility of an ongoing incident or secondary events. In these situations we need a feed at the scene showing a spreadsheet (table) that lists bed capacities per nearby hospitals. This table is shared by headquarters and updated. In such situations, casualties should be directed and escorted to the selected hospitals without delay and decontaminated while receiving the appropriate treatment.

Triage has to be resilient.

14.1.2 TRIAGE Is the Main Tool Helping Emergency Medical Services to Save and Care for People

Triage is a classification tool to categorize simultaneous casualties.

In a disaster situation, triage rules are simple, based on clinical symptoms and should not take time. FRs should save the maximum number of lives possible, if not all, and emergency medical services (EMS) should dispatch care accordingly and as early as possible. Different triage systems are currently in use. Some are based on colours and others on acronyms.

At an early stage, most triage rules rely on visual assessment and basic algorithms such as: the person walks, or can't walk; the person moves, or can't move; the person breathes, or isn't breathing. Later, the triage sort will use a scoring system using physiological parameters. In all cases medical data should be kept confidential.

The primary concern is how quickly and professionally initial triage is performed and how often triage is redone over time. Triage should be redone at various intervals to see if anything has changed or if inappropriate triage was not done properly and has to be redone (Frykberg 2002).

Triage is a continuous procedure with several check points: close to the epicentre of the event, before and after onsite treatment, at the entrance of the hospital. In the specific case of a CBRNE event, triage must also be performed before decontamination. The goal is to contribute to an efficient organization on site avoiding queues or overcrowding.

Triage coordinators should be medically highly qualified, trained to do triage, well considered and accepted as leaders. In France, medical pre hospital teams are used to work with other field partners on a daily basis and are professionally well considered.

Triage covers all types of casualties not only somatic but also psychological. Capacity to handle the psychological impact generated by the disaster situation should not be overlooked. Medical units are trained to provide people with early information on post-traumatic stress disorders and prevention measures. We must keep in mind that FRs, EMS and crisis managers are also psychologically affected.

14.1.3 Triage Is Not Only a Technical But Also an Ethical Matter

Ethical issues are a very important topic that should be part of all thinking and procedures at all stages of a disaster: planning, preparedness, operational procedures in particular the triage phase, and recovery.

International guidelines such as the World Medical Association's Declaration of Lisbon (1981, 1995) proclaim that medical care must not violate any universally applicable ethical standards. These guidelines help medical and health professionals to recognize ethical dilemmas in medical and health care and provide general rules and principles. This Lisbon declaration on ethics and emergencies highlights key points: preserve autonomy; offer the best health care, avoid negative consequences, preserve equity, prevent doctors to be under pressure.

As it is performed by human beings, triage is not only a technical matter but is also an ethical matter. It is not possible to eliminate the effect of the emotional factors upon dealing with a large number of victims. Studies have shown that such stress usually causes over-triage, which usually benefits the casualties (Hogan and Lairet 2002). These data establish the importance of triage accuracy and triage discrimination as a major determinant of casualty, minimizing both undertriage and overtriage (Frykberg 2005).

In France, Dr. Martinez, who was working in the Paris emergency department, developed a formula to perform triage and priority of care. This formula (Priority = S + T + C + V) comprises 4 parameters: the severity of the case (S), the delay in delivering care (T), the immediate care (C) which can be offered, and the social impact (V).

An unexpected mass event affecting health is always a social event involving children, elderly and vulnerable people. It can rapidly become a political event if the proposed solutions do not meet the expectations of the population. The wish of each person is to be taken care of and well protected without delay. This feeling is shared in particular by VIPs due to their social status. On a daily basis, these well-known personalities will immediately create media concerns and represent a social constraint to bear in mind when dealing with an emergency situation. Thus, in a CBRNE event, all people have to go through the same procedures such as decontamination. It is not possible to divert resources because VIPs want preferential treatment. Triage in the case of a disaster should consider this social factor in advance. Ethically speaking, the notion of favouritism should vanish in a collective event.

In the case of CBRNE mass casualties leading to mass vaccination, one may consider that health providers have the priority. This point has to be discussed and agreed on in advance in contingency plans. The smallpox plan highlights the measures necessary for managing the anti-epidemic campaign. The plan relies on the concept of "ring vaccination," in which identification of smallpox cases is followed by vaccination of health care providers who likely will be in contact with the affected individuals and will have to treat them.

14.1.4 Public Information

CBRNE incidents have a low probability. However, if an incident of this type occurs, it will have a high impact on the general public and result in anxiety. Ethical issues require transparent and timely information for the public. This issue requires special media training during the preparation phase. Indeed, it is important to give factual, trustworthy and timely information to the public and rescue services while sorting out victims. Procedures such as decontamination may be frightening and not well understood without previous information. Decontamination is preceded by some levels of disrobing and may be embarrassing for the public. These psychosocial aspects of mass decontamination including public communication needs are key factors to increase public compliance and fit the needs of vulnerable groups who may require increased assistance.

14.2 What Is Specific to CBRNE Terrorist Attack Situations?

Although there is no clear indication that CBRNE agents would be used by terrorists, the less we know about the agents, the greater the impact on our behaviour is.

CRNE incidents are more time critical and have a rapid onset of effect, thus, it is important that training and response plans can guarantee the efficiency of immediate countermeasures. Mr. Lagadec, international expert and former senior scientist at Ecole Polytechnique says: "We must be prepared for the unexpected event" (Lagadec 1991).

Among these countermeasures we must highlight the need of antidotes, thus the need to identify as quickly as possible the agent. Exposure to nerve agents and hydrogen cyanide may be lethal within minutes without the appropriate antidotes.

Another specific countermeasure is the need to assess the scene, cordon it in order to establish a safe distance from the location of the contaminant. Three CBRN functional zones are defined (AMedP7.1, NATO 2015):

- The hot zone, close to the point of release. This is a non-permissive area where there is a direct hazard (primary exposure and contamination) to the FR from the environment. It is also called the exclusion zone in the French terminology whereas for NATO this is a zone where despite protective measures, a significant risk to the responder remains that can only be reduced by avoidance (e.g. high dose radiation source or explosive device).
- The warm zone is a semi-permissive buffer area where there is a secondary contamination hazard due to contaminated equipment, personnel or casualties leaving the hot zone. A significant vapour hazard may remain. It is bounded by a cordon that is called the clean/dirty line (CDL). This is where appropriate stabilisation and decontamination of casualties are performed prior to their evacuation towards the cold zone.
- The cold zone is devoid of CBRN hazard and can be represented by hospitals (Fig. 14.1).

Hazard management includes protection of the FRs and decontamination.

Protection of first responders is mandatory and most of FRs are civilians. The selection of PPE is the responsibility of employers, in order to preserve FR's safety and health while performing their tasks and on site triage. It is also the employer's responsibility to train their employees to properly wear PPE during exercises. Note

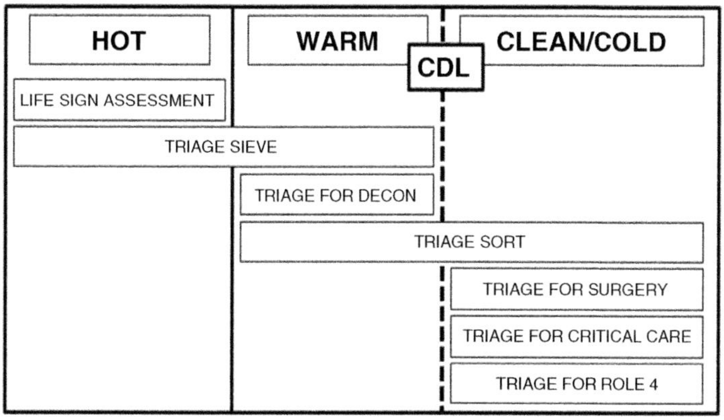

Fig. 14.1 Triage systems and CBRNE zones (from NATO AMedP-7.1)

that civilians are not trained on a daily basis and not as healthy as military personnel. That is why European projects such as the IFREACT PPE (FP7 EC Project: grant agreement no. 285034) look for PPE with a very low physiological burden, a negligible performance degradation and a high level protection factor. Decision-makers can take into consideration the advantage of the panoramic vision provided by the large visor of the IFREACT FRH16 overpressure hood in order to reassure victims and facilitate triage: FRs, in particular EMS see and can be seen by the casualties.

Serious injuries such as trauma or severe haemorrhages may lead to dilemmas. CBRNE incidents use the same categories as conventional incidents to ensure interoperability and ensure trauma, CBRNE and combined casualties have the same casualty flow priorities.

14.2.1 C Triage

At difference to most biological or radiological events, those produced by chemical agents may induce very quickly life-threatening situations that require rapid responses from the FRs. For agents creating a high risk of contamination transfer, specific steps should be taken. One of the challenges is also to quickly determine the agent in order to treat accordingly. Easy-to-use and fast triage tools are needed (Nyberg et al. 2011).

We follow current triage rules:

FRs, wearing PPE, evacuate casualties from the site of the event ("hot zone" or "exclusion zone") to the Casualty Collection Point (CCP) as early as possible,.

At the CCP, FRs collect the descriptions of victims' symptoms (tick on a list). It is best to consider all casualties entering the medical system as contaminated.

Indeed, EMS may not have the ability, or time, to positively rule out contamination. Indeed, for severe casualties time is of the essence.

At some distance from the "hot zone", a "warm zone" is delineated. Some vapour hazard may still be present and cross-contamination from the casualties is to be taken into account. Before decontamination, triage must take place in order to determine the priority to enter the decontamination chain. A generic all-hazards approach adapted from the management of trauma is now recommended by NATO for CBRNE casualty management including the management of trauma in a CBRNE environment. (AMedP7.1, NATO 2015) The priorities for treatment are CAaBCDE: **C**atastrophic haemorrhage, **A**irway and **a**ntidote administration (Medical countermeasure), **B**reathing and oxygen delivery, where appropriate, **C**irculation, **D**econtamination (and disability) and **E**vacuation to a more permissive environment.

In absence of combined injuries, nerve agent casualties are at high risk of being T1 while for the vesicants for instance most will be T2 (Delacour et al. 2014). NATO classifies casualties as T1 if immediate treatment is needed, T2 if treatment can be delayed, T3 for minimal treatment and T4 for expectant treatment.

During triage, it should be determined if the victim is poisoned or presents with conventional injuries. Indeed, chemical agents may be dispersed with explosive devices. The 1995 Tokyo sarin attack (without explosives) should not be considered as the only possible scenario considering the use of UAV as potential vectors of chemical agents.

One issue to be considered during decontamination is the fact that casualties will seek for their relatives. Ethical issues require keeping family members together hence the capacity of the system to identify and trace them. This issue should be taken into account in the response process thanks to tagging and tracing casualties, mentioning their relatives on the same tag. That is mandatory and of utmost importance in order to avoid separating family members.

Information to understand ongoing procedures should be given to families.

The tag and trace function is simple and based on different technologies, bearing in mind that its functionality must be in compliance with the communication system available at the site.

Another issue with decontamination procedures can arise if patients refuse to undress due to different cultural codes, and this should be taken into account and alternative procedures planned.

14.2.2 B Triage

Following a terrorist attack with biological agents, triage is specific because of the delay to symptoms after infection and the possible spread of cases with transmissible agents. The emerging infectious diseases like Severe Acute Respiratory Syndrome (SARS) and Pandemic Influenza have alerted national decision-makers to the need for contingency plans, trying to maximise outcomes from the available resources (e.g. antibiotics, antidotes, vaccines, ICU beds, respirators, dialysis machines, blood products).

14.2.3 Radiological and Nuclear Event Triage

A few years ago, we conducted an exercise in Paris with a scenario involving a radiation dispersal device ("dirty bomb"). We experienced that constraints due to PPE were almost negligible (unpublished results). Triage procedures were also facilitated by the use of detection devices (radioisotope identification using a portable gamma spectrometer).

Efficient radiological triage is the prerequisite to providing treatment to casualties that would develop acute radiation syndrome and therefore needs early specific medical support. The main challenge to be overcome is to bridge the gap between the diagnosis and categorization of casualties (unexposed individuals or worried well (1), exposed individuals not requiring specific medical treatment (2), and exposed individuals requiring early administration of hematopoietic growth factors/cytokines (3) which currently relies on irradiation dose estimation and the initiation of specific treatments for those in category 3 (Dörr et al. 2017). The initial assessment of radiological victims includes the timing and severity of prodromal clinical signs and symptoms. The prodromal phase is the earliest clinical phase of the Acute Radiation Syndrome (ARS) (roughly within the first 24 h after exposure to ionizing radiation). The recording of the onset (e.g. time to vomiting) and severity (e.g. gravity and frequency of vomiting) of clinical signs is very important for diagnosis and casualty categorization. This is complemented by serial complete blood counts with differential, repeated every 6–12 h. Three to four days are required to perform the Gold Standard cytogenetics-based biological dosimetry. The latter marker assesses the global radiation dose received but does not effectively differentiate total-body and partial-body exposures which require neither the same level nor the same earliness of medical support (Hérodin and Drouet 2005).

Recent progress has been made to improve/anticipate both the diagnosis of exposed individuals and the treatment of the hematopoietic syndrome of acute radiation sickness (Abend et al. 2016). After many years of extensive preclinical research, the Food and Drug Administration (FDA) approved in 2015 the use of granulocyte colony-stimulating factor (G-CSF or filgrastim, a cytokine stimulating factor that increases the production of neutrophil granulocytes) for mitigating radiation-induced neutropenia in nuclear/radiation accident victims (Vijay et al. 2015). The sooner G-CSF treatment starts the more effective it is (i.e. 24 h after irradiation is recommended), and therefore it is important to shorten the time required for diagnosis. Recent work strongly suggests that reliable diagnosis of external irradiation using the METREPOL (Medical Treatment Protocols for Radiation Accident Victims) clinical severity score (including hematopoietic, gastrointestinal, neurovascular and cutaneous sub-syndromes evaluation) could be achieved within 3 days of irradiation and even earlier.

From recent events we can highlight the need for transparent information because radiological and nuclear risks provoke more fears and worries in the public than other CBRNE events.

14.3 Conclusion

CBRNE events can cause thousands of victims requiring early medical intervention to avoid the development of health consequences. Triage is, consequently, a vital goal in all levels of contingency planning, both military and civilian. Site and personal safety is of paramount concern for the FRs. In case of radionuclide contamination, casualties should be tested for contamination as part of the initial triage; in case of chemical contamination, this assessment should not delay emergency medical procedures. Initial treatment will begin after local decontamination is performed. Medical management will be improved after thorough decontamination.

Few studies address the validation of triage procedures and tools in the case of CBRNE mass casualties (Culley and Svendsen 2014). The goal of European projects is to develop field trials on medical care and site management during triage in a CBRNE incident (EDEN, TOXI TRIAGE projects). The efficacy of triage relies on training and simulation in exercises.

References

Aacharya, R.P., C. Gastmans, and Y. Denier. 2011. Emergency department triage: An ethical analysis. *BMC Emergency Medicine* 11: 16. http://www.biomedcentral.com/1471-227X/11/16.

Abend, M., C. Badie, R. Quintens, D. Oskamp, S. Strunz, S. Moertl, S. Doucha-Senf, S. Dahlke, J. Menzel, and M. Port. 2016. Examining radiation-induced in vivo and in vitro gene expression changes of the peripheral blood in different laboratories for biodosimetry purposes: First RENEB gene expression study. *Radiation Research* 185 (2): 109–123. https://doi.org/10.1667/RR14221.1. Epub 2016 Feb 1.

AMedP-7.1.2015. NATO medical management of CBRN casualties. https://www.shape.nato.int/resources/site6362/. Brussels: NATO.

Culley, J.M., and E. Svendsen. 2014. A review of the literature on the validity of mass casualty triage systems with a focus on chemical exposures. *American Journal of Disaster Medicine* 9 (2): 137–150. https://doi.org/10.5055/ajdm.0150.

Delacour H., O. Nespoulous, C. Renard, F. Ceppa, and F. Dorandeu. 2014. Chapitre 21: Prise en charge des intoxications chimiques aiguës dans les structures de rôles 1 et 2. In *Blessé de guerre,* ed Stéphane Mérat, 415–428. Arnette, France

Dörr, H., M. Abend, W.F. Blakely, D.L. Bolduc, D. Boozer, T. Costeira, T. Dant, A. De Amicis, S. De Sanctis, M. Dondey, M. Drouet, F. Entine, S. Francois, G. Gagna, N. Guitard, F. Hérodin, M. Hoefer, A. Lamkowski, G. La Sala, F. Lista, P. Loiacono, M. Majewski, P. Martigne, D. Métivier, X. Michel, J. Pateux, J. Pejchal, G. Reeves, D. Riccobono, Z. Sinkorova, L. Soyez, D. Stricklin, A. Tichy, M. Valente, C.R. Woodruff Jr., L. Zarybnicka, and M. Port. 2017. Using clinical signs and symptoms for medical management of radiation casualties – 2015 NATO

exercise. *Radiation Research* 187 (3): 273–286. https://doi.org/10.1667/RR14619.1. Epub 2017 Feb 20.
Frykberg, E.R. 2002. Medical management of disasters and mass casualties from terrorist bombings: How can we cope? *The Journal of Trauma* 53: 201–212.
———. 2005. Triage principles and practice. *Scandinavian Journal of Surgery* 94: 272–278.
Hérodin, F., and M. Drouet. 2005. Cytokine-based treatment of accidentally irradiated victims and new approaches. *Experimental Hematology* 33 (10): 1071–1080.
Hogan, D.E., and J. Lairet. 2002. Triage. In *Disaster medicine*, ed. D.E. Hogan and J.L. Burstein, 10–15. Philadelphia: Lippincott Williams and Wilkins.
Lagadec, P. Mars. 1991. *La gestion des crises outils de réflexion à l'usage des décideurs*. Paris: Mc Graw-Hill.
Nyberg, A.G., D. Stricklin, and A. Sellström. 2011. Mass casualties and health care following the release of toxic chemicals or radioactive material-contribution of modern biotechnology. *International Journal of Environmental Research and Public Health* 8: 4521–4549. https://doi.org/10.3390/ijerph8124521.
Vijay, K. Singh, Patricia L.P. Romaine, and Thomas M. Seed. 2015. Medical countermeasures for radiation exposure and related injuries: Characterization of medicines, FDA-approval status and inclusion into the strategic National Stockpile. *Health Physics* 108 (6): 607–630.
World Medical Association Declaration on the Right of the Patient: Lisbon, 1981, 1995.

Chapter 15
Health Care Workers' Obligations in CBRNE Crises

Dónal P. O'Mathúna

Abstract Health care workers (HCWs) often suffer the brunt of injuries during chemical, biological radiological, nuclear and explosive (CBRNE) events. Throughout history, those caring for the injured, dying and dead put themselves at risk of harm, infection or contamination. Recent events include the 2014–2016 infectious outbreak of Ebola virus disease in West Africa and the targeting of health facilities in the conflict in Syria. Decisions by HCWs to care for others in the face of such risks have been lauded as heroic whether undertaken for personal moral reasons or in response to an ethical duty to care. However, some have questioned whether such a duty to care is ethically obligatory in the face of some CBRNE events. Ethical analysis of the SARS outbreak found that additional ethical reflection was needed on HCWs' obligations during CBRNE events. The ethical arguments used to justify the duty to care are reviewed in this chapter. However, other duties exist for HCWs which may conflict with the duty to care. The World Health Organization's guidance on ethics in pandemics notes that the duty to provide care in pandemics is not unlimited, and that employers and governments have reciprocal obligations to provide training and protective equipment to HCWs during CBRNE. Empirical research raises questions about whether health care organisations are adequately prepared for CBRNE, particularly for the ethical decision-making that will be required. Rather than taking a regulatory or legal approach to this issue, this chapter will argue that the ethical virtues of courage and volunteerism should be fostered in HCW training. In keeping with a virtue ethics approach, leadership takes on an important role in ethical decision-making, as well as praising those who respond to CBRNE by caring for others in spite of the personal risks and their conflicting obligations.

Keywords Duty to care · Health care workers (HCWs) · Ethical obligations · Courage · Virtue ethics · Leadership

D. P. O'Mathúna (✉)
School of Nursing and Human Sciences, Dublin City University, Glasnevin, Dublin 9, Ireland

College of Nursing, The Ohio State University, Columbus, OH, USA
e-mail: donal.omathuna@dcu.ie

D. P. O'Mathúna, I. de Miguel Beriain (eds.), *Ethics and Law for Chemical, Biological, Radiological, Nuclear & Explosive Crises*, The International Library of Ethics, Law and Technology 20, https://doi.org/10.1007/978-3-030-11977-5_15

15.1 Introduction

Health care workers (HCWs) have a crucial role to play in responses to chemical, biological, radiological, nuclear and explosive (CBRNE) events. These workers often take on serious risks when they do so. The Black Death (bubonic plague) swept through Medieval Europe on various occassions, leading to debate over whether physicians and other public officials had a duty to remain with the sick (Luther 1527). During the 1918 influenza pandemic, many nurses and others who volunteered to care for the infected themselves contracted the deadly influenza (Godderis and Rossiter 2013). After the 1995 Aum Shinrokyo Sarin attacks, 10% of hospital HCWs involved in treating exposed patients ended up needing to be treated themselves because patients' clothes exposed them to Sarin (Rebera and Rafalowski 2014). In the 2002–2003 SARS outbreak, approximately 30% of all cases occurred in HCWs; in Toronto, about half of the cases were in HCWs, three of whom died (Malm et al. 2008). Towards the end of the Ebola virus disease outbreak in West Africa, 898 cases of Ebola infection had occurred in HCWs, of whom 518 died (Statista 2015). The lethality of the virus is well-known, but the HCWs were disproportionately impacted. Liberia was worst hit, where 8.07% of its health workforce died from the virus, compared to 0.11% of its general population (Evans et al. 2015). Not only did caring for patients put HCWs at risk of physical disease and death, but some who survived faced stigmatisation and discrimination afterwards (OHCHR 2014).

Yet possibly the most egregious example of the risks HCWs find themselves exposed to has been in the ongoing war in Syria. In spite of the traditional and internationally protected status of health care facilities and workers during conflict and war, they have become common targets of violence (ICRC 2015). In Syria alone, between November 2015 and December 2016, over 400 violent acts against health care facilities have been documented (Elamein et al. 2017). These resulted in 677 injuries and 261 deaths, with one quarter of the victims being HCWs. As a result, health care services are being divided between locations and staff numbers on site are kept to a minimum to reduce the impact of individual attacks.

HCWs who respond to CBRNE events typically put themselves in harm's way to an increasing degree compared to responders to other disasters that do not involve infectious agents or contaminants. Disaster responders typically choose to enter disaster zones, and usually have specialised training to prepare for such events. This chapter will not focus on such responders, but on HCWs who find themselves in the midst of a CBRNE event. They may find themselves grappling with whether or not they have an ethical obligation to continue to provide health care to those injured in the event. The ethical question is whether HCWs have a duty to care for any and all patients in the event of a CBRNE crisis. In other words, are HCWs obliged to accept the risks that accompany caring for patients after a CBRNE event? In recent years, 'the issue of duty to care has emerged as a matter of paramount concern among health care professionals, hospital administrators, public policy makers, and bioethicists' (Ruderman et al. 2006, 2).

15.2 The Current Situation

Before proceeding to examine the ethical issues involved, this discussion must be prefaced by the finding that many HCWs accept the risks of their work and diligently care for their patients in spite of the dangers involved. The heroism apparent during the Ebola outbreak led *Time* to declare these HCWs the 2014 Person of the Year (Von Drehle and Baker 2014). A report into the SARS outbreak in Toronto found that, 'Workers generally showed heroism and altruism in the face of danger during the SARS outbreak' (University of Toronto 2005, 10). However, this was not the case for all. The Toronto report continued that 'some balked at caring for people infected with SARS, and a few were dismissed for failing to report for duty. Post-SARS, many health care workers raised concerns about the level of protection to themselves and their families. Some even left the profession' (University of Toronto 2005, 10).

CBRNE crises lead to ethical challenges that involve individual and societal decisions. On an individual level, workers have to balance their ethical duties towards their patients and the public with those towards themselves and their own health, and their duties towards their families and other dependants. The societal level is seen most clearly by the devastation brought by Ebola onto the healthcare workforce in the affected countries. These systems started with woefully inadequate healthcare workforces, and the deaths of so many HCWs from Ebola have left the systems even more debilitated. Although Ebola was directly devastating in many ways, an additional 25,000 deaths may occur each year because of the deaths of so many HCWs in these countries (Evans et al. 2015). A healthcare system, and society more broadly, must somehow balance its ethical duty to care for those presently sick while knowing that this can increase the risks to the long-term health of the people in their societies who will require care in the future.

With the constant risk of an incurable pathological agent appearing, the growing fear of an imminent pandemic, and the threat of terrorists obtaining nuclear materials, the importance of reflecting on the duty to care and other ethical challenges associated with CBRNE has been noted (and was one of the impetuses for the two projects than contributed to this book). Ethical decision-making skills are needed to address such issues. Yet in spite of the acknowledged importance of such topics, the Toronto SARS report found that 'There is currently a vacuum in this field' (University of Toronto 2005, 10). In the US, for example, the 2009 H1N1 flu pandemic led to several calls to develop pandemic readiness plans. Such plans identify the duty to care as one of several ethical principles (Koenig et al. 2011). Some US states have pandemic plans, but few address the duty to care explicitly. Louisiana, the state most directly hit by Hurricane Katrina, is one that does, defining the duty to care as 'the obligation of health care professionals to care for patients at all times' (Louisiana Department of Health & Hospitals 2014, 10). It notes that during pandemics HCWs will have to balance their duty to individual patients with that to all patients. Yet no discussion is provided about how HCWs should balance this duty with their duties to themselves and their families.

At the same time, other developments have occurred within healthcare that challenge the traditional view of this issue. Workers' rights and the importance of creating safe workplaces is increasingly recognised, appropriately so. Yet these developments could raise questions about policies that put HCWs at increased risk of harm. As noted above, the duty to care usually refers to individual patients. During a crisis, like a CBRNE event, a balance may be needed between the duty to care for individuals and for society as a whole. In additional, the duty not to harm is a strong ethical principle. HCWs may see their duty not to harm patients as obliging them to avoid contagious or contaminated patients because of the risk that they could pick up the agent and pass it on to others unknowingly. While the duty to care is a long-standing ethical principle in healthcare, it raises many ethical challenges that require further careful reflection and analysis.

15.3 Healthcare Readiness for CBRNE

Some discussion has occurred around the readiness of current healthcare systems for emergency responses to CBRNE. The World Health Organization (WHO) Regional Office for Europe published a checklist to assist hospital administrators and emergency managers prepare for disasters. The guidance states that 'Effective human resource management is essential to ensure adequate staff capacity and the continuity of operations during any incident that increases the demand for human resources' (WHO Regional Office for Europe 2011, 17). However, the document does not discuss the ethical issues faced by HCWs, nor does it provide guidance on ethical decision-making. Instead, the focus is on policy and management preparation.

A systematic review examined qualitative research into nurses' preparations for ethical issues in public health emergencies and disasters. The authors identified 'a failure to directly address the issue of ethical considerations in planning, preparedness, and response to public health emergencies and disasters by nurses' (Johnstone and Turale 2014, 73). The American Nurses Association (2017) acknowledged the conflict between nurses' duty to care for patients and their own right to self-preservation, and the challenge of finding the right balance. The American College of Chest Physicians (CHEST) has a prominent role in preparing US doctors who will be at the front line of any respiratory pandemic. They issued a Consensus Statement on the ethical issues of caring for patients during pandemics and disasters. This Statement does not address the duty to care (Biddison et al. 2014).

With little explicit emphasis on the duty to care in crisis situations, HCWs may not be aware of their obligations or have reflected on how they might respond during a CBRNE crisis. Some surveys have asked HCWs if they would report for work in the event of different types of crises. In a survey in the US, roughly half of the HCWs who responded stated they likely would not report to work during an influenza pandemic (Balicer et al. 2006). Another survey of Emergency Department workers in Chicago asked whether people would work additional hours to help victims of different types of CBRNE events. Among these HCWs, 98% said they would accept extra work after an airplane crash, 85% after a radioactive bomb, and 54% after the release of a biological agent (Masterson et al. 2009).

These types of studies have weaknesses because they ask people's opinions about hypothetical situations when they are not in a disaster. Just as questionnaires about past events suffer from recall bias, questionnaires about the future can also be biased. Studies of actual behaviour over a number of decades have found that the vast majority of those in emergency roles fulfilled their duty when called (Scanlon 2014). However, the on-going nature of a pandemic and the lethality of agents like Ebola, sarin and radioactivity, may lead some to question whether their duty to care applies to those patients.

Changes to the nature of healthcare are also important to consider here. In some countries, healthcare is increasing viewed and practiced according to a business model. Arguments are made to view healthcare organisations as businesses, patients as consumers, and HCWs as employees contracted to deliver a service. As such approaches come to infuse the ethos of healthcare, they may inadvertently impact responses to CBRNE crises. Sheri Fink's investigation into Hurricane Katrina included an exploration of some of the corporate decision-making that interferred with patient care and rescue (Fink 2013). This sort of approach can lead to a focus on contractual obligations towards workers and patients, which may not address obligations to society as a whole. As will be discussed more below, a contractual approach has particular limitations when altruism and sacrifice are required, which is usually the situation when the duty to care is invoked.

15.4 The Traditional View

Granted the limitations of the surveys cited above, the possibility that a significant proportion of the healthcare workforce might not show up in the aftermath of a CBRNE crisis is of concern. This conflicts with what can be called the traditional view of the duty to care in emergencies. Daniel Defoe captured this view dramatically in his fictionalised account of the Great Plague of London which may have claimed up to 20% of the population in 1665. The book examines many of the same ethical issues that challenged the responders to Ebola virus disease.

> So the Plague defied all Medicine; the very Physicians were seized with it, … This was the Case of several Physicians, even some of them the most eminent; … it rather is to their Praise, that they ventured their Lives so far as even to lose them in the Service of Mankind; They endeavoured to do good, and to save the Lives of others (Defoe 1969, 35–6).

This traditional view is that doctors and nurses will lay down their lives for their patients and the good of society. This ethic has been lived out in HCWs and disaster responders throughout history. Those who stayed to take care of Ebola patients, or came to offer what they could, acted upon this view (Von Drehle 2014). In battle-torn Syria, the White Helmets formed a volunteer group to provide aid and rescue to those injured by chemical and explosive devices. Their motto is 'Whoever saves one life, saves all of humanity' (Malsin 2016, 23). Among responders like these, the duty to care focuses on putting the needs of others above ones own needs. The belief is that those who have specialised training, or more opportunities than others, thereby have a responsibility to help those in need. Martin Luther gave similar advice to those facing the Black Death: 'paid public servants such as city physicians, city clerks and constables, or whatever their titles, should not flee unless they furnish capable substitutes who are acceptable to their employer' (Luther 1527, 477).

15.5 Changing Professional Ethics

This view was explicitly called for in earlier professional ethics codes. The 1847 *Code of Medial Ethics* of the American Medical Association stated that physicians have a 'duty to face the danger, and to continue their labours for the alleviation of the suffering, even at the jeopardy of their own lives' (American Medical Association 1847, 27). Something very similar was included in the 1922 Code of Ethics of the Canadian Medical Association (Ruderman et al. 2006). But things have changed. The 1949 International Code of Medical Ethics of the World Medical Association stated that, 'A doctor must give emergency care as a humanitarian duty unless he is assured that others are willing and able to give such care' (World Medical Association 1949). By the 1970s such strong statements about the duty to care in the face of infectious risks had disappeared from most codes (Ruderman et al. 2006).

The traditional obligation that a doctor should accept the risks of practicing medicine has faded. Different reasons have been given for this, including the general belief that infectious diseases had been overpowered, as least in those countries with access to effective antibiotics (Ruderman et al. 2006). The move from healthcare as a calling to one where healthcare is a rewarding and finantially comfortable career has been implicated in this change, as well as the business model for healthcare organisations. The arrival in the 1980s of HIV and AIDS, especially at first when little was known about its transmission and less about its treatment, led to further weakening of the meaning of the duty to care. SARS, H1N1 and Ebola raised further questions about what the duty to care implies, both ethically and practically. In response, nurses in Liberia went on strike during the Ebola outbreak because of low wages and the lack of personal protective equipment (PPE) that guards against infectious diseases like Ebola (Agence France-Presse 2014). Likewise, nurses in the US protested, and some went on strike, to express concern over the lack of preparation and protective equipment for Ebola in their hospitals (Skinner and Johnson 2014).

In the 1800s, a doctor knew that caring for some patients entailed serious risks. In some parts of the world, that never changed. Many HCWs have died, and continue to die, in the service of their patients. As one doctor noted during Ebola, many HCWs 'found themselves "fighting a forest fire with spray bottles." They did not give up' (Von Drehle 2014). The prospect of a pandemic has brought things full circle and requires much more extensive discussions about the ethical requirements entailed in the duty to care.

15.6 Ethical Justification

The arguments brought up to ethically justify the duty to care can be grouped into five approaches, each with its counter-arguments. They will be summarised here, but are discussed in more detail with an extensive bibliography by Malm et al. (2008).

The first is that HCWs consent to take on certain risks as they go through their professional training. Risks are part of all forms of practice, and can arise physically, such as from needle-sticks, using certain equipment, lifting patients, etc., and also emotionally, such as from watching patients suffer and even die. Training should make people aware of these risks and therefore HCWs consent to accept these risks as part of professional practice. This is especially the case when someone works in infectious diseases, but applies to all HCWs. The criticism of this argument is that SARS, H1N1 and Ebola have changed healthcare practice. Risks from serious infectious diseases did not exist for many HCWs trained in earlier years, and hence people have not in any realistic way consented to take on some of the risks that exist today.

The second approach uses an analogy to argue that HCW give implied consent to certain risks. If someone becomes a soldier, she cannot legitimately object to the risks of combat. If someone becomes a firefighter, he cannot object to getting close to fires. As part of the decision to pursue particular careers, certain risks must be accepted even if this is done implicitly or changes over time. The criticism of this argument is that even if accepted in principle, it does not necessarily entail that all HCWs have a duty to care for all patients, particularly those who put HCWs at higher risk. For example, someone with training in infectious diseases may have a duty to care for infected patients, but this does not imply that a psychiatric nurse has the same duty to those patients. Thus, the duty to care should be limited to those who have relevant training and experience in the conditions that ail the patient.

The third argument is that someone with specialised knowledge, training and skills has an additional responsibility to use that knowledge to help those in need of that expertise. By their own training, HCWs are better able to care for such patients in ways that reduce their own risks. They also have easier access to the necessary protective equipment. However, objections to this approach point out that HCWs differ widely in their training and skills. If this implies that HCWs have different duties to care, this will lead to a confusing range of obligations.

The fourth approach uses that of reciprocity, or a social contract view. Many HCWs receive assistance from society in various ways (through scholarships or research funds) and often receive privileges and status from their careers, not to mention good salaries. As a result of such contributions from society, HCWs should reciprocate by taking care of patients who are in need. Some object that HCWs receive widely varying contributions from society and also vary widely in the privileges they receive. Some claim that the respect or status once given HCWs no longer exists, and in some places HCWs do not receive a commensurate salary.

The fifth approach is that many HCWs take professional oaths or accept ethics codes and these often commit them to a duty to care. However, these declarations tend to be very general and also change with time. People entering a professional may view oaths as symbolic rather than a serious commitment to specific ethical values. Further, a review of 61 professional codes found that 85% either did not mention the duty to care or offered no clear guidance on its implementation (Upshur et al. 2006).

One further approach is regularly mentioned by those who volunteer to serve: the importance of serving one's community. The White Helmets help their fellow Syrians because, according to their director, 'At the end of the day, this is my country' (Malsin 2016, 26). The Ebola fighters regularly mentioned that they wanted to help their neighbours and their communities (Von Drehle 2014). This aspect gets at the importance of the internal ethical motivations of HCWs, and in particular the ethical virtues that will be discussed below. Healthcare ethics in Western contexts has become focused on individuals and their rights. Public health crises and CBRNE events remind us of the need to consider relationships and communities within ethics. At the same time, such approaches do not provide clear mechanisms for balancing the duties HCWs have to the variety of people they are in relationships with: family, neighbours, colleagues, patients. Guidance is needed on how to balance such conflicting obligations to various parties.

15.7 WHO Guidance

The World Health Organization (WHO) developed general guidance on the ethical issues in public health responses to pandemic influenza (WHO 2007). The document devoted one section to the duty to care, and provided some guidelines. This acknowledged that there is a need for open discussions and agreement between HCWs, their professional organisations, and the public on the duty to care. What was stated after SARS remains the case today: 'the time to address the ethical duty to provide care is at hand – before the arrival of the next public health emergency' (Ruderman et al. 2006, 6).

The guidance notes that the duty to care can be addressed in terms of moral, professional, contractual or legal obligations. A strong case is made for approaching this as a moral obligation. At the same time, WHO acknowledges that 'the duty to work notwithstanding risks to one's own health is not unlimited' (WHO 2007, 14). Rather than taking a rigid legalistic approach, the guidance offers some flexibility. Specific policies about the duty to care should be developed within jurisdictions through dialogue and consultation with all stakeholders. These should take account of differences in expertise and skills possessed by various HCWs, although as needs develop, people may be asked to work beyond their usual responsibilities. The risks that HCWs are asked to accept should reasonably be expected to make a difference in the pandemic. This highlights the importance of policies being based on the best available evidence. Reasonable accomodations should also be made for those whose own health changes their risk, such as those who may be immunocompromised or pregnant. A mechanism should be available by which risks are distributed among individuals and groups in an equitable, fair and transparent way. Open consultation and dialogue help to show whether policies are viewed as fair and just.

As HCWs accept additional risks, employers and governments have reciprocal obligations to reduce risks to HCWs. These include providing training for pandemics, implementing appropriate preventive measures, and having available the necessary equipment to respond adequately, such as personal protective equipment (PPE). When people become ill or injured through their work, treatment should be provided, as well as access to psychosocial support. To uphold the duty to not harm, employers and governments should educate HCWs on their ethical obligation to reduce the spread of infection if they become ill.

These recommendations are applicable for HCWs responding to many CBRNE events. However, they were published in 2007 and their uptake has been relatively slow. For example, a CBRNE readiness survey was sent to all hospitals in Belgium and the results published in 2014. Seventy two percent of the hospitals responded, with 11% stating they had decontamination facilities close to the emergency department entrance, and 6% reported having appropriate PPE available for those doing triage and decontamination. At the same time, almost three-quarters of the facilities expressed the belief that they were ready for CBRNE events. The researchers concluded that 'There are serious gaps in hospital preparedness for CBRN incidents in Belgium' (Mortelmans et al. 2014, 300).

Much further work is needed to engage with HCWs on the duty to treat and the practical obligations this entails. Policies should be developed based on discussions between all parties involved, including the public. At the same time, evidence needs to be collected to support related policy and practice. As changes are made, and as CBRNE events occur, data should be collected on the effectiveness of various approaches, policies and recommendations, including those related to ethical issues. This area thus overlaps with chapters in this book on research ethics so that research conducted to generate evidence is carried out in ethically appropriate ways.

15.8 Ethical Virtues and the Duty to Care

While codes and regulations have their place, they have limitations. In his study of trust in the helping professions, Edmund Pellegrino finds that contractual agreements tend to lead to 'ethical minimalism' (1991, 79). 'The professional's necessity to efface self-interest will be blunted since legalistic and contractual relationships call upon the participants to protect their own interest, not that of the other party—except to the extent the contract requires. The impetus to do the "extra" that requires some compromise of self-interest is blunted if not destroyed entirely' (ibid.). In CBRNE crises, going the extra mile is exactly what needs to be encouraged, and this requires a return to traditional professional values and what are called ethical virtues: the personal character traits that lead people to strive for ethical ideals.

When the duty to care is seen as based on ethical virtues, as opposed to on legal or contractual obligations, HCWs will be guided by their personal integrity and consciences on how to balance this duty with their other duties. Concerns have been raised that this may lead to many HCWs failing to report for work during CBRNE events. A small amount of research has been conducted on the factors that motivate people to work during CBRNE events. In one study, the leading motivational factor was a sense of duty to their profession, but other factors included concern for their family's health, personal safety and child care (Masterson et al. 2009). These are very personal factors, rather than ones based on professional and legal obligations.

When interviewed for *Time's* 2014 Person of the Year award, Ebola virus disease responders did not refer to regulatory or professional codes. Instead, they were motivated by personal factors. 'Ask what drove them and some talk about God; some about country; some about the instinct to run into the fire, not away.' Some were inspired by those who had come from far away. Some survived Ebola and said 'It looked like God gave me a second chance to help others' (Gibbs 2014).

Ways of communicating these personal ethical virtues need to be developed and included in the training of HCWs, especially those more likely to be involved in CBRNE events. The traditional way to encourage virtuous development has been through narratives (O'Mathúna 2008). Stories about courage, heroism and fear in the face of life's dangers have been used throughout human history to encourage reflection on the sorts of character traits that are necessary to deal with life's risks. Such was the purpose of the ancient folk tales and more recent moral fables. The real-life stories of responders such as those recounted in the Ebola issue of *Time* can be used to foster the development of virtues like courage and volunteerism in the face of CBRNE risks (Von Drehle 2014).

Leadership and mentoring are other important factors in this area. Dr. Jerry Brown was the Medical Director & General Surgeon of ELWA Hospital in Monrovia, Liberia. As he helped set up an early Ebola treatment unit (ETU) he realised that his leadership and example were going to speak more loudly to the other HCWs than any policy, code or contract. 'He was now forced to … suit up in Tyvek and go to work in the ETU. Every willing hand was needed, and the fearful staff must see that the boss had enough courage to do as much as he asked of them' (Von Drehle 2014).

Leadership by example is more compatible with virtue ethics than leadership by decree or contract. Those leading facilities and developing policies may at times need to take on the same risks as other front-line workers. Dr. Carlo Urbani accepted this as part of his role within the WHO and as a result became exposed to SARS and died from the infection (Ruderman et al. 2006). His leadership and example can motivate others to have the moral courage to serve in the face of CBRNE risks.

15.9 Conclusion

Stories set in dramatic circumstances sometimes receive much publicity. In this way, the movie *Megan Leavy* tells the story of a US Marine and her dog Rex whose job was to search for explosives to protect their colleagues (Cowperthwaite 2017). In spite of the dangers to themselves, she and other CBRNE responders enact the virtues needed to fulfil the duty to care in crisis situations. The *Time* stories thus reveal modern-day heroes who chose to act on their duty to care as a result of their virtues and in spite of their fears. It is just as important to praise those who in less dramatic ways act on the basis of their moral courage and exemplify the ethical traits needed to promote sacrificial behaviour on behalf of others and society. These are ordinary people who do extraordinary things based on virtues like courage, compassion and commitment to the good of others. As those values and virtues are promoted, the duty to care will once more be seen as part of what it means to be a virtuous healthcare worker.

References

Agence France-Presse. 2014. Ebola: Liberian nurses strike over lack of protective equipment. *The guardian*, September 2. http://www.theguardian.com/world/2014/sep/02/ebola-liberian-nurses-strike-lack-protective-equipment. Accessed 28 Nov 2017.

American Medical Association. 1847. *Code of medical ethics of the american medical association*. Chicago: American Medical Association Press. https://collections.nlm.nih.gov/catalog/nlm:nlmuid-63310410R-bk. Accessed 28 Nov 2017.

American Nurses Association. 2017. *Who will be there? Ethics, the law, and a nurse's duty to respond in a disaster*. https://www.nursingworld.org/~4af058/globalassets/docs/ana/ethics/who-will-be-there_disaster-preparedness_2017.pdf. Accessed 28 Nov 2017.

Balicer, Ran D., Saad B. Omer, Daniel J. Barnett, and George S. Everly Jr. 2006. Local public health workers' perceptions toward responding to an influenza pandemic. *BMC Public Health* 6 (99): 1–8.

Biddison, Lee Daugherty, Kenneth A. Berkowitz, Brooke Courtney, Marla J. De Jong, Asha V. Devereaux, Niranjan Kissoon, Beth E. Roxland, Charles L. Sprung, Jeffrey R. Dichter, Michael D. Christian, and Tia Powell. 2014. Ethical considerations. Care of the critically ill and injured during pandemics and disasters: CHEST consensus statement. *Chest* 146 (4 Suppl): e145S–e155S.

Cowperthwaite, Gabriela. 2017. *Megan leavy*. Los Angeles: LD Entertainment.

Defoe, Daniel. 1969. *A journal of the plague year*. Oxford: Oxford University Press.

Elamein, Mohamed, Hilary Bower, Camilo Valderrama, Daher Zedan, Hazem Rihawi, Khaled Almilaji, Mohammed Abdelhafeez, Nabil Tabbal, Naser Almhawish, Sophie Maes, and Alaa AbouZeid. 2017. Attacks against health care in Syria, 2015–16: Results from a real-time reporting tool. *Lancet* 390 (10109): 2278–2286.

Evans, David K., Markus Goldstein, and Anna Popova. 2015. Health-care worker mortality and the legacy of the Ebola epidemic. *Lancet Global Health* 3 (8): e439–e440.

Fink, Sheri. 2013. *Five days at memorial: Life and death in a storm-ravaged hospital*. New York: Crown.

Gibbs, Nancy. 2014. The choice. *Time*, 10 December. http://time.com/time-person-of-the-year-ebola-fighters-choice/. Accessed 28 Nov 2017.

Godderis, Rebecca, and Kate Rossiter. 2013. 'If you have a soul, you will volunteer at once': Gendered expectations if duty to care during pandemics. *Sociology of Health and Illness* 35 (2): 304–308.

ICRC (International Committee of the Red Cross). 2015. *Health Care in Danger: Violent Incidents Affecting the Delivery of Health Care, January 2012 to December 2014*. https://www.icrc.org/en/publication/4237-health-care-danger-violent-incidents-affecting-delivery-health-care-january-2012. Accessed 28 November 2017.

Johnstone, Megan-Jane, and Sue Turale. 2014. Nurses' experiences of ethical preparedness for public health emergencies and healthcare disasters: A systematic review of qualitative evidence. *Nursing and Health Sciences* 16: 67–77.

Koenig, Kristi L., Hoon Chin Steven Lim, and Shin-Han Tsai. 2011. Crisis standard of care: Refocusing health care goals during catastrophic disasters and emergencies. *Journal of Experimental and Clinical Medicine* 3 (4): 159–165.

Louisiana Department of Health & Hospitals. 2014. State hospital crisis standard of care guidelines in disasters. http://www.phe.gov/coi/Documents/LA%20State%20Hospital%20CSC%20Guidelines%20in%20Disasters.pdf. Accessed 28 Nov 2017.

Luther, Martin. 1527. Whether one may flee from a deadly plague. In *Martin Luther's basic theological writings*, 2012, ed. Timothy F. Lull and William R. Russel, 3rd edn, 475–487. Minneapolis: Augsburg Fortress.

Malm, Heidi, Thomas May, Leslie P. Francis, Saad B. Omer, Daniel A. Salmon, and Robert Hood. 2008. Ethics, pandemics, and the duty to treat. *American Journal of Bioethics* 8 (8): 4–19.

Malsin, Jared. 2016. The white helmets of Syria. *Time*, October 16: 21–26. http://time.com/syria-white-helmets/. Accessed 28 Nov 2017.

Masterson, Lori, Christel Steffen, Michael Brin, Mary Frances Kordick, and Steve Christos. 2009. Willingness to respond: Of emergency department personnel and their predicted participation in mass casualty terrorist events. *Journal of Emergency Medicine* 36 (1): 43–49.

Mortelmans, Luc J.M., Sam Van Boxstael, Harald G. De Cauwer, and Marc B. Sabbe. 2014. Preparedness of Belgian civil hospitals for chemical, biological, radiation, and nuclear incidents: Are we there yet? *European Journal of Emergency Medicine* 21 (4): 296–300.

O'Mathúna, Dónal P. 2008. Teaching ethics using popular songs: Feeling and thinking. *Monash Bioethics Review* 27 (1–2): 42–55.

OHCHR (Office of the High Commissioner for Human Rights – West Africa Regional Office. 2014. A human rights perspective into the Ebola outbreak. http://www.globalhealth.org/wp-content/uploads/A-human-rights-perspective-into-the-Ebola-outbreak.pdf. Accessed 28 Nov 2017.

Pellegrino, Edmund D. 1991. Trust and distrust in professional ethics. In *Ethics, trust, and the professions: Philosophical and cultural aspects*, ed. Edmund D. Pellegrino, Robert M. Veatch, and John P. Langan, 69–89. Washington, DC: Georgetown University Press.

Rebera, Andrew P., and Chaim Rafalowski. 2014. On the spot ethical decision-making in CBRN (Chemical, biological, radiological or nuclear event) response. *Science and Engineering Ethics* 20 (3): 735–752.

Ruderman, Carly, C. Shawn Tracy, Cécile M. Bensimon, Mark Bernstein, Laura Hawryluck, Randi Zlotnik Shaul, and Ross E.G. Upshur. 2006. On pandemics and the duty to care: Whose duty? Who cares? *BMC Medical Ethics* 7 (5): 1–6.

Scanlon, Joseph. 2014. Ethical issues in health communications: Strategies for the (inevitable) next pandemic. In *Disaster bioethics: Normative issues when nothing is normal*, ed. Dónal P. O'Mathúna, Bert Gordijn, and Mike Clarke, 95–106. Dordrecht: Springer.

Skinner, Curtis, and Kia Johnson. 2014. U.S. nurses hold strikes, protests over Ebola measures. *Reuters*, 12 November. http://www.reuters.com/article/2014/11/13/us-usa-nurses-ebola-idUSKCN0IW14S20141113. Accessed 28 Nov 2017.

Statista. 2015. Ebola cases and deaths among health care workers due to the outbreaks in West African countries as of November 4, 2015. https://www.statista.com/statistics/325347/west-africa-ebola-cases-and-deaths-among-health-care-workers/. Accessed 28 Nov 2017.

University of Toronto Joint Centre for Bioethics Pandemic Influenza Working Group. *Stand on guard for Thee. Ethical considerations in preparedness planning for pandemic influenza.* 2005:10. http://jcb.utoronto.ca/publications/documents/stand_on_guard.pdf. Accessed 28 Nov 2017.

Upshur, Ross, Tarif Bakdash, Daniel Fu-Chang Tsa, James Johnson, and Susan Maybud. 2006. The role and obligations of healthcare workers during a pandemic influenza outbreak. WHO working draft. http://www.who.int/ethics/PI_Ethics_draft_paper_WG3_14Sept06.pdf. Accessed 28 Nov 2017.

Von Drehle, David, with Aryn Baker. 2014. The Ebola fighters: The ones who answered the call. *Time*, 10 December. http://time.com/time-person-of-the-year-ebola-fighters/. Accessed 28 Nov 2017.

WHO. 2007. Ethical considerations in developing a public health response to pandemic influenza. http://www.who.int/csr/resources/publications/WHO_CDS_EPR_GIP_2007_2/en. Accessed 28 Nov 2017.

WHO Regional Office for Europe. 2011. Hospital emergency response checklist. An all-hazards tool for hospital administrators and emergency managers. http://www.euro.who.int/en/health-topics/emergencies/disaster-preparedness-and-response/publications/2011/hospital-emergency-response-checklist. Accessed 28 Nov 2017.

World Medical Association. 1949. International code of medical ethics. http://www.cirp.org/library/ethics/intlcode/. Accessed 28 Nov 2017.

Chapter 16
Criminal Protection of Public Health in the Field of Biological Emergencies: The Case of the Spanish Legal System

Emilio José Armaza-Armaza

Abstract This paper analyses the suitability of the Spanish legal criminal system to deal with the hypothetical case in which a person subjected to a biosecurity isolation regime (that is, to "quarantine" as he is a carrier of a serious illness) decides to ignore the advice of the health authorities (or, if applicable, the Judge), escaping from the hospital facilities where he is deprived of his liberty (we will begin describing two similar and real examples of this situation, one of them occurred in the African context, and the other one in Spain). In addition to the identification and analysis of rights or interests at stake (life, individual health, public health, freedom of movement, etc.), we propose the creation of a criminal offense to overcome the legislative loophole that we will identify which is also present in the criminal systems of most European and Latin American countries.

Keywords Public health · Quarantine · Biosecurity · Diseases · Epidemics · Bioterrorism

16.1 Diseases and Epidemics as a Source of Relevant Risk for Criminal Justice

On 20 July 2014, a Liberian lawyer who worked as a top official of the Ministry of Economy of his country, travelled to neighbouring Nigeria, specifically to its capital, the city of Lagos (Shuaib et al. 2014). At that time, some West African countries (including Liberia, Guinea and Sierra Leone) were struggling with the largest Ebola outbreak in their recent history. However, and this fact should be emphasised, at the

E. J. Armaza-Armaza (✉)
University of Deusto, Bilbao, Spain

Chair in Law and the Human Genome Research Group at the University of the Basque Country, Leioa, Spain
e-mail: emilio.armaza@deusto.es

D. P. O'Mathúna, I. de Miguel Beriain (eds.), *Ethics and Law for Chemical, Biological, Radiological, Nuclear & Explosive Crises*, The International Library of Ethics, Law and Technology 20, https://doi.org/10.1007/978-3-030-11977-5_16

time of the events described, not a single case of contagion of Ebola in Nigeria was reported.

This incident would not have been significant, had it not been that the official in question was subsequently referred to as "patient zero" (the index case) of the introduction of Ebola virus disease into Nigeria (Shuaib et al. 2014). Indeed, the first fears, which increased when Lagos health care workers were required to attend to the passenger who had collapsed some minutes after landing, were confirmed a few hours later when, already at the hospital, some specific tests were performed and confirmed the fact that the Liberian official was a carrier of the deadly virus.

Shortly after being informed in detail of his critical situation, the official expressed to the health personnel attending him, his total rejection of the health care treatments offered at the health centre, as well as his wish to leave the hospital facilities at that time. Evidently, the hospital authorities, taking into account the type of illness that would eventually be exposed to the inhabitants of the city of Lagos and, with all likelihood, the inhabitants of all Nigeria,[1] ignored the requests of the Liberian official and, on the contrary, ordered his isolation in restricted areas suited for this purpose. However, the response of the aforementioned official was not long in coming. Extremely aggravated and enraged, he extracted the cannulas connected to his body, undressed and urinated on the health staff attending him, so as to force them to stand back so that he could escape from the hospital facilities where he was being restrained (Osanyin 2015).

We must point out that the incident in question did not end there. After this altercation, the official was able to phone the Liberian Ambassador to Nigeria, and inform him of the details of his critical situation. The Ambassador's response was astonishing: He contacted the hospital authorities where his fellow citizen was isolated to inform them that he was going to file a complaint concerning the illegal detention given that, with their behaviour, they were violating the official's "Diplomatic immunity" (Osanyin 2015). He also outlined the serious consequences that this incident would bring to the political, economic and social relations established between the two countries.

To conclude this brief story, we should point out that despite the enormous political pressure exacted on the hospital authorities where the patient was isolated, the response was always the same. This person had to be subjected to a quarantine regime until the risks for public health derived from Ebola were ruled out. In the following hours, to effectively contain the carrier of this deadly disease, several hospital workers had to restrain the individual to prevent him from escaping, which put at risk the health and life of many people.

This Liberian official died 4 days after this incident (specifically on 24 July). He was not the only person who died that year in Lagos due to this terrible disease; four members of the medical staff who physically restrained the Liberian official were infected with Ebola died, in addition to other civilians exposed to the virus indirectly. It should be noted that other persons were isolated in quarantine as they had

[1] To this we can add that this behaviour also put in danger of contagion the inhabitants of the countries with which Nigeria maintains terrestrial or air connections.

been directly exposed to the virus, although fortunately none of them finally suffered the disease.

Now, we must remember that, although with some important differences, a similar incident occurred in Spain and placed our population in a hazardous situation that, perhaps, should not be downplayed. Indeed, the now-famous nurse who survived an infection with the Ebola virus, but when she was a carrier of the deadly virus, she went to a beauty centre located in Alcorcón (Madrid) in order to remove facial hair, and also sat for an exam at the Complutense University of Madrid (being very close to hundreds of people) (Rego and Ortíz 2014). In this sense, it is difficult to identify, as well as to explain, the reasons why this nurse had no objection to perform the activities listed in this paragraph, even when she consciously decided to adopt certain precautionary measures inside her home in order to prevent transmitting the disease to her husband (both decided to sleep in different beds, as well as use different bathrooms).

These examples – one from international experience, and the other one from the most well-known biological incident in Spain – may perhaps teach us that, in the context of a biological crisis, our society may have to deal with the conduct of certain people who refuse to be subjected to an isolation regime for health or biosecurity reasons (that is, what is commonly referred to as "quarantine") or who, without refusing to comply with a measure of this nature, consciously adopt behaviours that may endanger public health. The motivations for this type of behaviour, of course, could be diverse. Some people, who are extremely scared, will simply want to go back home; while others are carrying out a terrorist activity (Sánchez Moreno and García Estévez 2013). Others could act with the intention of transmiting the disease to other persons in order to pressure the Government to fund biomedical research or for the medical treatment of such disease.

In any case, what we would like to emphasise is that, regardless of the motivation of this hypothetical person, we would have to face the type of behaviours that endanger the "individual" health and life of the people who are finally infected, as well as the "public" health of a group of citizens particularly exposed (in a "specific" or "abstract" way) to one specific disease. However, from the perspective of Criminal Law, we are also interested in analysing the suitability of our criminal legislation to prevent and, where appropriate, sanction this type of behaviour.

16.2 The Isolation Order for Reasons of Biosecurity (Quarantine)

Before initiating the analysis of the role played by criminal law in the context of the "breaking of a quarantine order", it is necessary to ask if in our legal system there is a basis that allows the Public Administration to restrict the ambulatory freedom of citizens, with the pretext that consists in affirming that there is a risk that one person may be a carrier of a disease that could endanger the life or health of other citizens.

A first approximation to this question will necessarily lead us to examine the provisions of Article 25.3 of the Constitution (CE78), which clearly states that "The civil administration may not impose sanctions that, directly or indirectly, involving imprisonment." As we are aware, this precept has its origin in the need to draw the limits of the competencies of the criminal and administrative bodies, limits that, prior to EC78 were extremely diffuse and that in many cases resulted in Administrative sanctions that, in addition to constituting deprivation of liberty, would even be more rigorous than the criminal sanction. In any case, we must emphasise that the constitutional precept referred to exclusively points to the prohibition that the Administration has to impose "sanctions", which directly or indirectly imply deprivation of liberty. However, as we know, the notion of "sanction" has, of course, been thoroughly studied both in the Criminal Law Framework and in the seat of Administrative Sanction Law. Without a detailed analysis of this question, it should be pointed out that in the studies carried out within both spheres of legal knowledge, the idea that one of the main characteristics of administrative and penal sanctions is constituted precisely by the idea of "retribution". In this way, both types of legal consequences are configured as mechanisms of a clear retributive nature, in the sense that they allow the reaffirmation of the legal system that has been previously broken, as well as the atonement of guilt by those transgressors who voluntarily accept the sanction as a just consequence for the offence committed.

In any case, what should be made clear is that any restriction on an individual's ambulatory freedom due to the risk that he may be carrying a disease that could endanger the life or health of other citizens is not, in any way, a penalty. The realisation that such person is a carrier of a contagious and serious disease (such as Ebola) is not part of the catalogue of cases (criminal or administrative offences) that enable either the Criminal Law or the Sanctioning Administrative Law to impose a specific penalty. We are, therefore, faced with an administrative measure that restricts or limits the right to outpatient freedom, but which, by its complete lack of any type of content of a retributive nature (which leads us to the conclusion that it cannot be considered a sanction), could be perfectly imposed by the Administration, without violating the constitutional mandate mentioned above.

There are two legislative sources in Spain's legal system that regulate the cases in which isolation measures can be adopted for biosafety reasons. The first of these is in the *Organic Law 3/1986, of 14 April, of Special Measures in the Field of Public Health*. Article 2 of that regulation states that "The competent health authorities may adopt measures for the recognition, treatment, hospitalization or control when there are reasonable indications that there is a hazard to the health of the population due to a specific health situation of a person or group of persons or by the health conditions in which an activity is carried out." Likewise, in accordance with article 3 of the same Organic Law, "In order to control communicable diseases, the health authority, in addition to carrying out general preventive actions, may adopt appropriate measures for the control of the sick, for persons who are or have been in contact with them and the immediate environment, as well as those deemed necessary in case of risk of a transmissible nature." On the other hand, it is interesting to note that *Law 29/1998, of 13 July, regulating Contentious-Administrative*

Jurisdiction, establishes in its article 8.6 that "(...) it shall correspond to the Courts of Contentious-administrative the Judicial authorization or ratification of the measures that the health authorities consider urgent and necessary for public health and imply deprivation or restriction of freedom or another fundamental right."

We are, therefore, confronted with a measure envisioned in Spain's legal system whose justification lies, as it could not be otherwise, with the need to avoid the production of a major evil, starting from a "consequentialist" analysis of the hypothetical variants of the situation (Singer 2003). A situation analogous to the criminal figure known as the "justifiable state of necessity" is thus formed, by which, we recall, that the State allows a determined actor to sacrifice a lesser legal right – the ambulatory freedom of a specific person, for example – in order to safeguard another or more valuable right – the life or personal integrity of several citizens.

However, what has been said so far leads us to the conclusion that, according to our dogmatic tradition and, of course, to the provisions of our legal system, it is clearly acceptable and, of course, feasible to impose, even coercively, a restriction on the ambulatory freedom of an individual who is presumed to be suffering from a serious illness and who may thereby endanger the life or health of his fellow citizens. In other words, we have envisaged the possibility of sacrificing individual rights (among them, in particular, the right to freedom of movement) to protect public health against high-risk situations.

16.3 Criminal Law Before the Cases of Rupture of an Order of Isolation for Reasons of Biosecurity

Once we have identified "public health" as the fundamental value that justifies the sacrifice of ambulatory freedom for reasons of biosecurity, through the imposition of a quarantine measure, it is necessary to ask the following question: Does our criminal system have certain legal tools designed to punish those who succeed in evading the measure of isolation that has been legitimately imposed on them, either by fleeing from the place where they were already held or by avoiding being detained before they have been isolated?

16.3.1 A Legally Protected Collective Value in Danger: Public Health

The answer to the question given in the previous paragraph is dependent, once again, on the identification of the legally protected value that has been harmed or that, where appropriate, has been specifically or abstractly endangered due to violating the quarantine measure. This, of course, to ensure respect for the harm and offence principle, (fundamental to the Criminal Law) by which, we recall, it is dependent, enabling criminal intervention prior to injury or, at least, danger of injury to one of the criminally relevant legal rights (Boldova Pasamar 2016). The Harm Principle must allow democratic States to criminalise that conduct that threatens, as well as conduct that causes, harm. Indeed, if our interest is to prevent effective harms, we have good reasons to criminalise those behaviours that create a serious risk of harm to others.

So, what is the legal value that we would have tried to protect in this type of situation? Recall that we are facing that hypothetical situation in which a person carrying a serious and infectious disease breaks the order placing that person under quarantine. In this sense, we must first differentiate the criminal responses that would apply to the hypothetical offender when the crime of effective harm has been committed (for having, of course, produced a given result – such as the death or actual injury of one or more persons). In such cases, the criminal regulation for murder, homicideor injury will be applicable (all of which, probably, qualified by the production of a serious result). Given that these regulations aim to protect two legal values of an individual nature, the life and individual health of persons who have effectively suffered harm from such legal values.

However, this pathway is not the ideal route to protect the fundamental value that, as we have seen in previous paragraphs, has served to enforce the ethical and legal legitimacy of the application of the isolation or quarantine measure. We refer, particularly, to the idea of "public health", understood as something beyond individual health as the health of the community or, to be clear, the sum of the health of several persons (Romeo Malanda 2016).[2] Indeed, there is no doubt that the behaviour exhibited in the hypothetical situation raised as a *de facto* case in this paper, is particularly suitable to fall into an "effective" and "real" danger of harm to that legally protected value. In this sense, it serves as a quick empirical analysis of the most emblematic cases of recent biological crises caused by fatal diseases to realise that, even in those situations in which no person was effectively injured, there are multiple and even unimaginable risks for the health of the community.

16.3.2 Criminal Protection of Public Health in the Spanish Legal System

For the above reasons, we consider it appropriate to make a panoramic reading of the Penal Code (PC) articles that contain crimes against public health,[3] to identify the type of criminal that would enable the person who endangers the health of the

[2] It should also be pointed out that criminal protection of the aforementioned legal right is merely a reflection of the constitutional tutelage that has been established for this purpose (Article 43.2 EC states that "It is the responsibility of the public authorities to organize and protect Public health through preventive measures and the necessary services. The law shall establish the rights and duties of all concerned").

[3] All of them incorporated in Chapter III of Title XVII of Book II of the PC.

community through the violation of a quarantine measure. In this sense, we can indicate that the articles incorporated in the section of the PC devoted to the sanctioning of crimes that endanger public health can be grouped into the following blocks: (1) Crimes related to substances harmful to health or chemical products which can cause havoc. (2) Pharmacological crimes. (3) Crimes related to sports doping. (4) Crimes related to products of mass consumption. (5) Crimes related to drug trafficking.

The first of these groups of infringements is made up of the criminal provisions incorporated into articles 359 and 360 of the PC, in which the process, dispatch, supply or trade of the indicated substances or chemicals is sanctioned under certain conditions, processing, dispatch, supply or trade of the indicated substances and chemical products. The second group, consisting of the crimes listed in Articles 361, 362, 362 second, 362 and 362 fourth of the PC, is dedicated to punishing the so-called "pharmacological crimes", by means of which it punishes, also under certain circumstances, (processing, production, manufacture, import, export, marketing, etc.), which have in common the material object of the offense (medicineand other medical devices). Meanwhile, the third group consists of the basic form, as well as the aggravated assumptions, of a doping offense in the field of sport (Article 362 fifth of the PC) punishing, among other behaviours, the prescription, dispensation, supply, offering, etc. of prohibited substances or methods to athletes (professional or not) who wish to increase their physical abilities or modify the results of the competitions in which they participate (Atienza Macías and Armaza-Armaza 2016). The fourth group of crimes (Articles 363, 364 and 365 of the CP) is typified by a series of behaviours which affect the quality of products for human consumption and that, in general terms, with the exception of specific figures, have in common the material object of the crime (food products). Finally, in the fifth group of offences (typified in articles 368 et seq.), it punishes the basic and attenuated type and the aggravated forms of trafficking in toxic drugs, narcotic drugs or psychotropic substances, which specifically sanctions the possession and trafficking of materials that can be used for the processing of the substances mentioned above (precursors).

In short, this rapid review of the current wording of our Penal Code allows us to affirm that based on our system of criminal protection of public health (and, of course, the principle of legality), we could not sanction an individual who, being consciously a bearer of a pandemic and extremely lethal disease, endangers the health of the community through the violation of a quarantine measure. The behaviour is atypical (*Nullum crimen, nulla poena sine praevia lege*), including those cases – even more reprehensible – where the infected person intended to propagate and transmit the disease to others (Muñoz Conde and García Arán 2015). It seems, of course, a great paradox that our criminal legislator was not concerned about the development of tools for public health crimes that would prevent the type of behaviour being analysed here, even though it has special interest in the repression of conduct that could have scarce (or even null) suitability to jeopardize the legal good in question (for example, in SAP Valencia 306/2015 of 20 April, where the sale of 1.9 grams of hashish, with a purity of 12.5%, is a crime against public health, and is punishable with a penalty of 7 months' imprisonment).

16.3.3 Criminal Tools for Protection against Biological Weapons

However, without prejudice to the lack of health protection of the community in the face of these particular forms of threats, it is necessary to indicate that in our Penal Code we can find other penal precepts that, to a greater or lesser extent, are related to biological crises. In fact, this is the case with the crimes of:

1. Production of biological weapons (typified in article 160.1 of the PC and that, according to the majority doctrine, it is aimed at providing protection for the survival of the human species) (Romeo Casabona 2009; Armaza-Armaza 2015).
2. Development, manufacture, marketing, trafficking, deposit or use of biological, warfare, chemical, nuclear or radiological weapons or anti-personnel mines or cluster munitions (the modalities of which are typified in articles 566.1.1and 566.2 of the PC which, according to the majority opinion, are directed at providing protection to public order understood as well as tranquillity or peace in the collective manifestations of community life or as collective or community security) (Urruela Mora 2016).
3. Initiation of military preparation for the use of biological weapons, as well as the non-destruction of these weapons in violation of the international treaties or conventions to which Spain is a party (criminal modalities typified in article 566.2 of the PC and which are also directed at the protection of public order).

In any case, we must highlight the fact that all the crimes listed in the previous paragraphs are directed at the protection of the legal good that is not clearly identified with the fundamental value that serves as the basis for the legitimation of the imposition of a measure of isolation for reasons of biosafety; that is, they are not clearly identified with the protection of health in the community.

16.3.4 The Breach of Conviction and Serious Disobedience to Authority

Finally, it should be pointed out that the configuration of any of the offenses for parole violation (typified in articles 468 et seq. of the PC) could be discussed, but we should bear in mind that such precepts only sanction those who break a legal consequence of a criminal nature (Muñoz Conde 2015). In this way, the conduct of a person who violates an isolation measure for biological risk (which, we may recall, lacks connotations of a criminal nature) could not be subsumed in any of the types of penalties listed without sacrificing the criminal order (Article 4.1 CP), which prohibits the application of criminal precepts to cases other than those expressly included in them (*prohibition of analogy In malam departem*) (Boldova Pasamar 2016; Muñoz Conde and García Arán 2015). However, it is worth mentioning that, apparently, the only way that it could eventually lead to the punishment of the conduct analysed is opened by means of the crime of serious disobedience to the authority, its agents or even the private security personnel that develops private security activities in cooperation and under the command of the Forces and Security Corps (typified in article 556.1 of the PC) (Flores Mendoza 2016).

However, it is important to point out that in the first case (breaching a sentence), and in the second of these cases (disobedience to authority), the protected legal right has no relation to public health protection, since in the first case it protects the interest of the State in proper compliance with judicial decisions that impose penalties, security measures or precautionary measures (Flores Mendoza 2016). In the second, the so-called "principle of authority" is protected, and understood as the correct exercise of the functions that public powers exercise at the service of the community (Urruela Mora 2016).

16.4 Final Thoughts

At this point, the first issue to be highlighted is related to the fact that at present we do not have a criminal tool designed to punish those who know they are carrying a serious (and, of course, contagious) disease, and do not take appropriate measures to avoid endangering public health, or violate measures previously imposed on them by the administration (quarantine order). In light of the very serious dangers which may arise from biological crises and, in particular, the danger to the health of a community which can be triggered within the framework of the failure of a quarantine measure, at first sight there do not appear to be important disadvantages to including a new type of criminal measure that punishes those who adopt this type of behaviour.

However, it is extremely important to state that a new proposal for the "expansion of criminal law" must be accompanied by a profound debate on the justification and, of course, legitimation of this eventual process of primary criminalization, mainly because, as we are aware, the application of Criminal Law supposes, for those who suffer its legal consequences, the deprivation or impairment of particularly important rights and interests. In this regard, we believe that it is not only convenient but also essential to analyse this possible process of criminalization in the light of the fundamental principles of criminal law, to avoid the inclusion of a precept incompatible with the guarantees of the Rule of law.

Thus, it is important to analyse the implications that the principles of essentiality or fragmentarity may have for this matter (indeed, if this hypothetical "legal criminal" conflict can be effectively resolved through other forms of formal social control – less costly at a fundamental level. It is likely that we should choose the use of these, instead of resorting to the crudest tool available to us), of adverse effects (indeed, qualitative and quantitative limits must be established. The boundaries that should draw the line that separates an injury – or risk of injury – with criminal relevance, of an injury – or risk of injury – lacking the same), public interest and corresponding to reality (to develop a penal system that, although it can be labelled as "symbolic", is effective when individual opportunities for its application are available), certainty and specificity (to develop tools that do not affect legal certainty), guilt (to prevent the repression of people that might not be criticized for their conduct – thus respecting the principle of unjustifiable obedience to the law) and, finally, the principles related to the orientation of the purposes of the sentence (establishment of proportional and human penalties).

Acknowledgments Project framework "Bioterrorism and Biosafety: Bases for the structuring of new legal-criminal instruments to face biological threats" [www.bioterr.eu], Ref. DER2014-56634-JIN, funded by the Ministry of Economy and Competitiveness (MINECO) and developed at the University of the Basque Country UPV / EHU. This project is led by Dr. Emilio José Armaza-Armaza (Postdoctoral Researcher at the University of the Basque Country UPV/EHU – Professor of Criminal Law at the University of Deusto) and involved the participation of Dr. Carlos Maria Romeo-Casabona (Professor of Criminal Law at the University of the Basque Country UPV/EHU), Dr. Fernando Guanarteme Sanchez-Lazaro (Accredited Professor of Criminal Law at the University of La Laguna), Dr. Maria Angeles Cuadrado-Ruiz (Full Professor of Criminal Law of the University of Granada) and Dr. Fatima Flores-Mendoza (Professor of Criminal Law at the University of La Laguna).

References

Armaza-Armaza, Emilio José. 2015. Los instrumentos jurídico-penales para lucha contra el bioterrorismo. In *Bioterrorismo y Bioseguridad*, ed. Carlos María Romeo Casabona, 205. Bilbao: Cátedra Interuniversitaria de Derecho y Genoma Humano.

Atienza Macías, Elena, and Emilio José Armaza-Armaza. 2016. *El dopaje en el Derecho Deportivo actual: análisis y revisión bibliográfica*. Madrid: Editorial Reus.

Boldova Pasamar, Miguel Ángel. 2016. Los principios del Derecho Penal. In *Derecho Penal. Parte General. Introducción. Teoría Jurídica del Delito*, ed. Carlos María Romeo Casabona, Esteban Sola Reche, and Miguel Ángel Boldova Pasamar, 47–48. Granada: Editorial Comares.

Flores Mendoza, Fátima. 2016. Delitos contra la administración de justicia. In *Derecho Penal. Parte General. Introducción. Teoría Jurídica del Delito*, ed. Carlos María Romeo Casabona, Esteban Sola Reche, and Miguel Ángel Boldova Pasamar, 752. Granada: Editorial Comares.

Muñoz Conde, Francisco. 2015. *Derecho Penal. Parte Especial. 20ª edición, completamente revisada y puesta al día conforme a las Leyes Orgánicas 1/2015 y 2/2015, de 30 de marzo*. Valencia: Tirant lo Blanch.

Muñoz Conde, Francisco, and Mercedes García Arán. 2015. *Derecho Penal. Parte General. 9ª edición, revisada y puesta al día conforme a las Leyes Orgánicas 1/2015 y 2/2015, de 30 de marzo*. Valencia: Tirant lo Blanch.

Osanyin, Olaolu. 2015. The ebola patient migration – The Nigerian medicolegal perspective. *21st annual world Congress on medical law*.

Rego, Paco., and Ana M. Ortíz. 2014. Crónica: En la habitación del ébola. El Mundo archive. http://www.elmundo.es/cronica/2014/10/12/5438f3deca474167608b456f.html. Accessed 8 Oct 2018.

Romeo Malanda, Sergio. 2016. Delitos contra la seguridad colectiva II. Delitos contra la salud pública. In *Derecho Penal. Parte General. Introducción. Teoría Jurídica del Delito*, ed. Carlos María Romeo Casabona, Esteban Sola Reche, and Miguel Ángel Boldova Pasamar, 594. Granada: Editorial Comares.

Romeo Casabona, Carlos María. 2009. *Genética, Biotecnología y Ciencias Penales*. Bogotá: Ediciones de la Pontificia Universidad Javeriana – Grupo Editorial Ibañez.

Sánchez Moreno, Manuel, and Gustavo García Estévez. 2013. La amenaza bioterrorista procedente del yijadismo. In *Bioseguridad, Derecho y Defensa*, ed. María Ángeles Cuadrado Ruiz and Antonio Peña Freire, 166–170. Granada: Editorial Universidad de Granada.

Shuaib, Faisal, Rajni Gunnala, Emmanuel O. Musa, Frank J. Mahoney, Olukayode Oguntimehin, Patrick M. Nguku, Sara Beysolow Nyanti, Nancy Knight, Nasir Sani Gwarzo, Oni Idigbe, Abdulsalam Nasidi, and John F. Vertefeuille. 2014. Ebola virus disease outbreak – Nigeria, July–September 2014. *Morbidity and Mortality Weekly Report* 63 (39): 867–868.

Singer, Peter. 2003. *Ética práctica, Segunda edición*, 6–11. Madrid: Cambridge University Press.

Urruela Mora, Asier. 2016. Delitos contra el orden público I. Sedición. Atentados contra la autoridad, sus agentes y los funcionarios públicos, resistencia y desobediencia. Desordenes públicos. Tenencia, tráfico y depósito de armas, municiones o explosivos. In *Derecho Penal. Parte General. Introducción. Teoría Jurídica del Delito*, ed. Carlos María Romeo Casabona, Esteban Sola Reche, and Miguel Ángel Boldova Pasamar, 811. Granada: Editorial Comares.

Chapter 17
When Risk Management Systems 'Fail': On Criminal Negligence and the Limits of Scientists' Responsibility

Andrea Perin

Abstract This chapter consists of a brief discussion on some legal aspects concerning scientists' responsibility in risk prevention processes. After proposing some introductory considerations on scientists' responsibility as such, the author deals with the L'Aquila earthquake crisis of 2009, when a strong quake destroyed significant parts of L'Aquila (Italy) and surrounding villages, killing more than 300 people. The chapter focuses on the relations between scientific knowledge, normative expectations, decision-making and criminal negligence for 'failed' risk assessment and management, paying particular attention to the role of 'regulatory science' in constructing the 'reasonable person' normative standard of care in the theory of criminal negligence. This allows explaining why the first judgement in the L'Aquila trial (2012) is not convincing, having misunderstood how policy-relevant science should participate in prevention processes and the construction of normative standards. In his conclusions, the author suggests some reasons for the recent tendency to blame experts when natural or technological disasters occur.

Keywords Criminal negligence · Reasonable person · Normative expectations · Regulatory science · Precautionary principle · Risk management · L'Aquila case

17.1 The Responsibility of the Scientist

Since this book also deals with technological crises such as nuclear disasters, let us reflect on *science* and *responsibility*, starting from the historical and literary figure of Ettore Majorana (1906–1938, presumed death). Majorana, a hidden character in

---

A. Perin (✉)
Facultad de Derecho, Universidad Andres Bello, Santiago, Chile

University of Trento, Trento, Italy

University of the Basque Country, Greater Bilbao, Spain
e-mail: andrea.perin@unab.cl

D. P. O'Mathúna, I. de Miguel Beriain (eds.), *Ethics and Law for Chemical, Biological, Radiological, Nuclear & Explosive Crises*, The International Library of Ethics, Law and Technology 20, https://doi.org/10.1007/978-3-030-11977-5_17

the history of nuclear energy, was not only an extraordinary scientist but also probably the best Italian physicist of the twentieth century. As Enrico Fermi himself stated, he was 'a genius like Galileo and Newton' (Recami 2014: 408). Nonetheless, the importance of his works[1] started to be comprehended nearly 20 years after his mysterious disappearance.

The great Sicilian writer Leonardo Sciascia put forward the hypothesis that Majorana probably realised before anyone else that some of his and other physicists' works had arguably facilitated the development of the atomic bomb. Such a shocking thought might explain why on the 25 March 1938, he suddenly took the night boat to Palermo, leaving Naples (where he was living and teaching) and disappeared without a trace. Sciascia argued that Majorana had decided to abandon his identity and renounce his scientist status as soon as he foresaw the possible catastrophic consequences of his research (Sciascia 1987).

As Sciascia himself admitted, this fascinating hypothesis represents more a 'mix of history and invention' than a real historical explanation (Recami 2014: 408). Nonetheless, it maintains a strong evocative significance on a speculative level. In fact, in Sciascia's meaningful essay (Sciascia 1987), Ettore Majorana becomes a narrative character and a symbol that reminds us of the *vexata quaestio* concerning the responsibility of scientists towards society (Recami 2014). What are the limits of their ethical and juridical responsibility? For instance, should we[2] blame them for 'bad' foreseeable consequences of their activity or any harmful event they could have foreseen? If so, what does 'foreseeable' mean in juridical contexts? How can we evaluate (1) whether something was foreseeable in certain concrete circumstances and (2) how one should have acted in order to prevent the foreseeable event?

In his essay, Sciascia did not try to give answers (not explicitly, at least). Nonetheless, he finely reduced the 'is/ought' chasm, nimbly moving from 'mere facts' to a narrative 'model-of-conduct'. Indeed, he provided us with a bright portrayal of *his* 'good scientist', which is a scientist who takes responsibility and renounces his status as a scientist when his work could be misused by political powers (Recami 2014: 408).

Although some philosophers discuss the ethical distinction between 'acting' (which entails 'responsibility for one's actions') and mere 'doing' ('which concerns only the effective execution of a 'job', without concern for the wider consequences') in the so-called 'age of technology' (Galimberti 2009: 3), those questions concerning scientists' responsibility are undoubtedly still at stake (Douglas 2003). Since science and technology may have profound, destructive or irreversible effects, scientists should take responsibility for the technological applications that they make possible – intentionally or not – as well as for their possible consequences (Jonas 1984).

[1] He did not publish much in terms of quantity. He published nine articles before his disappearance (Recami 2006) and a tenth article, whose manuscript was found by Majorana's brother among his files, was published in 1942 in the Italian journal *Scientia*, now available also in English in *Quantitative Finance* (Mantegna 2006).

[2] The term 'we' might refer to anyone (members of society, academics, judges, institutions of law enforcement, etc.).

Nevertheless, scientists' responsibility does not only concern techno-science, i.e. what scientists do, just because scientific knowledge and new technologies enable them to do it. In fact, their supposed ability to *explain* and thereby *predict* events governed by natural laws might also justify their moral as well as strictly legal responsibility for what they do not prevent, in spite of their legal duty to (try to) do so.[3] Thus, due to their *scientific-technical skills*, scientists might be responsible for not acting 'properly' whenever their hypothetical (alternative) careful conduct could prevent harmful events.

Yet to what extent? We may agree with those who affirm that 'science is our best hope to face the dangers that threat both our planet and ourselves' (Harris 2014: 107–108). Nonetheless, we would find it rather questionable that, for instance, such an optimistic *portrayal of science* could justify the recently increasing involvement of scientists in risk management and prevention processes (and therefore their part of responsibility in them). Let us think: should scientists personally decide how people *ought to behave* when facing technological and natural risks? Should they establish *how to respond* facing crisis? If so, since actions and decisions entail moral, legal, in some cases even criminal liability for their consequences, we might infer that scientists should be held responsible whenever they are involved in a prevention system which 'fails' by not preventing the type of event that it is supposed to prevent.

Such a simplistic conclusion would be, again, rather questionable. By asserting that a major risk organisation shall correctly assess risks and successfully prevent disasters, we would actually support our hopes on *normative* (highly discretional and therefore controversial) *expectations*, rather than on *empirical evidence*. So the first question is the following: how do courts 'construct' normative standards of care? Moreover, we should consider that risk prevention processes consist of complex procedures composed of many possible stages: empirical observation (data collection), risk assessment (application of certain scientific knowledge to observed data), risk communication (report), risk management (policy-making and decision: how we ought to behave) and prevention (action). So, the second question at stake is: what are the potential roles and tasks of scientists in these procedures?

17.2 The Role of 'Regulatory Science' in Constructing the 'Reasonable Person' Normative Standard of Care in the Theory of Criminal Negligence

Courts might legally consider technological and natural disasters as ascribable to *someone's fault* or to *inevitable misfortune*. Indeed, almost all normative theories of criminal negligence – and therefore legal systems – divide adverse events (damages,

[3] For instance, according to the Italian Criminal Code (Article 40.2), "*whoever, against a legal duty of preventing an event, does not prevent it, shall be liable for its commission*" (translated by the author).

harms, disasters, catastrophes, and so on) into these two – or other similar and equivalent – main categories (e.g. Castronuovo 2009; Lauta 2015).

First of all, for any reader who may be unfamiliar with these concepts, we shall clarify in general terms what the 'fault/misfortune' divide means. On the one hand, the term 'misfortune' indicates events that we possibly can (scientifically) *explain* but not (legally) *ascribe* to someone's responsibility. This judgement might depend on the unforeseeability and/or unpreventability of these events, according to certain normative criteria and expectations. On the other hand, using the term 'fault' we usually imply that someone is culpable for what occurred. In particular, if courts estimate that a *reasonably prudent person* – e.g. a 'reasonable scientist' – could have foreseen and prevented the event that the defendant materially caused or did not prevent (having the legal duty to do so) he/she might be found 'negligent'. Thus, whoever *causes or does not prevent* harmful events (even though materially provoked by external forces of nature) *breaking normative expectations* can be accused and blamed.

Now, how could we 'construct' normative expectations, whose violation we define as 'negligence'? The issue of the limits of negligence is traditionally one of the most controversial in criminal law, especially with regard to the method used to define the objective and concrete 'duty of care' in the absence of statutory standards. In fact, when legal systems do not provide the courts with specific and concrete standards of diligence, judges have to determine their content, turning to the so-called 'reasonable person' normative standard (e.g. Holmes 1881; Marinucci 1965; Fletcher 1971; Moran 2003; Westen 2008; Baron 2012). In particular, when defendants belong to certain social groups, whose typical members are supposed to own *certain scientific-technical skills* – e.g. physicians, engineers, seismologists, etc. – scientific knowledge helps judges evaluate whether the concrete harmful event – e.g. a disaster – was *objectively foreseeable.*[4] If so, the 'foreseeability' normative criterion enables the interpreter to define the 'duty of care' indicating how the supposed 'prudent person' would have acted – i.e. how the defendant ought to have acted – in order to prevent the offence under the concrete circumstances (Forti 1990).

Hence, 'negligence' means *violation of a 'standard of care'*, that is a norm of conduct that has to be defined assuming *what a typical reasonable person, belonging to the social or professional group of the real perpetrator, would have done to prevent a specific, foreseeable harmful event under certain concrete circumstances.* Therefore, legal norms are not necessarily 'explicit, mandatory rule[s] of behaviour formally established by the state' (Bicchieri 2014: 209). They can be normative patterns ruled by a court, according to certain (highly discretional social-normative) criteria, and correspond to 'social norms' (Scalet 2003).

In order to clarify this point, we shall consider a paradigmatic (and frequent) case of medical malpractice. Imagine, for instance, that a medical practitioner fails in detecting the reason for certain signs and symptoms, and that his/her patient dies. In such a case, in order to decide whether the doctor is liable or not, the judge should

[4] See, in this regard, footnote 11.

comply with the following fundamental steps. First, according to the 'but-for' test and the 'abductive' method of reasoning (e.g. Tuzet 2006; Pardo and Allen 2008; Bex and Walton 2012; Douven 2017), he shall verify whether the undiagnosed disease was the effective cause of the event. Secondly, he shall question whether the scientific-technical knowledge of the expected 'reasonable doctor' would have been adequate for foreseeing the fatal outcome promptly and correctly. If so, he shall finally define how the defendant should have behaved. Consequently, the concrete 'duty of care' consists of how the hypothetical 'reasonable practitioner' would have acted according to his/her required scientific-technical skills, under the same circumstances.[5]

The same judicial reasoning should guide the courts in dealing with any case of failed risk prevention. Good case studies might be the 1963 Vajont dam disaster, the 1976 Seveso chemical explosion, the 1986 catastrophic nuclear accident at Chernobyl and the 1988 Clapham Junction railway accident, as well as more recent natural-technological crises such as the 2005 New Orleans flood and the 2011 Fukushima Daiichi nuclear disaster (Forti 1990; Lauta 2015; Chernov and Sornette 2016). Concerning all these cases, the main questions were (and still are) the following: did risk assessors and managers appeal to – and comply with – the proper *available*, *accessible*, and *reliable* scientific knowledge? Could this have enabled them to operate carefully and thereby prevent those tragic outcomes?

Judges' consideration of these events as mere *misfortunes* or otherwise *injustices*, then, mainly depends on how they face these issues, i.e. on how they *choose* the 'due' scientific-technological knowledge, according to the 'reasonable person' normative parameter, thereby deciding *whether* the event which occurred was objectively foreseeable, and *how* it was possibly preventable.

17.3 The L'Aquila Case

Now we shall start to address the second question mentioned in the introductory paragraph, observing what might happen when scientists are directly involved in the risk management stage. This happened, for instance, during the L'Aquila earthquake crisis in 2009, when after several weeks of foreshocks, a strong quake

---

[5] Let us specify that according to the so-called theory of 'objective imputation' – *objektive Zurechnungslehre* –, in order to fairly impute a harmful event to one's conduct, judges should also verify the existence of certain specific 'relations of risk' between the supposed negligent conduct and the resulting event (e.g. Roxin 1994; Gimbernat Ordeig 1985; Forti 1990; Donini 2006; Perin 2018). Moreover, in order to maintain the existence of *criminal* negligence, judges should finally evaluate whether the defendant can be possibly excused ('individual culpability'), for being subjectively unable to comply with the duty of care (e.g. Fletcher 1971; Westen 2008; Castronuovo 2009), or otherwise turning to the 'principle of excusability' – *Zumutbarkeit* – (e.g. Fornasari 1990; Melendo Pardos 2002). However, for the purposes of this chapter, we shall focus only on the judicial reasoning oriented to the construction of the objective 'duty of care', which appears to be similar – although not identical – in Common Law and Civil Law traditions (Fletcher 1971).

destroyed significant parts of L'Aquila (a medieval town in the Abruzzo region, in central Italy) and of the surrounding villages, killing more than 300 people. Again, for any reader who may not remember the case, we shall summarise the facts preceding the disaster.

On March 31, just a few days before the earthquake (which occurred on 6 April 2009), a meeting of some expert members of the 'National Commission for the Prevention of Major Risks' – an advisory body of the Department of Civil Protection on technical-scientific matters – took place in L'Aquila.[6] The National Civil Protection officially turned to them in order to carry out a serious analysis based on scientific evidence about the risk for the security of citizens, and thereby provide the agency with useful scientific information in order to manage the crisis. Nevertheless, in spite of this official target, a public official who attended the meeting and some of the expert attendees participated in a sort of 'press campaign' launched by the Chief of the Civil Protection, meant to reassure the population. In fact, since a local laboratory technician (who was not actually an expert in earthquakes) had made unofficial predictions based on measurements of radon gas levels and publicly declared that a catastrophe was imminent, the citizenship was understandably frightened (Koketsu and Oki 2015).

Thus, after the meeting, local media started to broadcast those committee members' declarations, recorded both before and during the meeting. For example, one of the six expert attendees, Enzo Boschi,[7] was asked during the meeting if the current seismic swarm could be a precursor to a major quake like the one that levelled L'Aquila in 1703. He answered: 'It is unlikely that an earthquake like the one in 1703 could occur in a short term, but this possibility cannot be totally excluded' (Amato and Galadini 2015: 191). Instead, interviewed before the meeting by local television, the public official who attended the meeting, Bernardo de Bernardinis,[8] announced: 'the scientific community continues to assure me that (…) it's a favourable situation (…), because of the continuous discharge of energy (…) the situation looks favourable'.[9] None of the scientists present at the meeting disproved this statement (not publicly, at least).

In the following days, many people who were living and sleeping on the street during the previous few weeks, because of the incessant foreshocks, decided to go back to their houses. There, many of them died, buried in the rubble, killed by the earthquake that took place only 6 days after the meeting and the 'press campaign'.

---

[6] We must highlight that the committee that met on March 31 was not the National Commission for the Prevention of Major Risks, because its 21 official members had not been invited. Only some of them were present, together with other parties. This is of crucial importance, because it entails that it is not possible to hold the six accused expert participants liable on the basis of the regulation that provides the duties (of care) for the National Commission members.

[7] Former president of Italy's National Institute of Geophysics and Volcanology (INGV).

[8] Former vice-president of the Civil Protection Agency.

[9] This declaration, recorded before the meeting, is available on line: https://www.youtube.com/watch?v=kLIMHe0NnW8. Accessed 9 Oct. 2018.

Could anyone have done something to prevent the tragedy? Yes, at least according to the Public Prosecutor of L'Aquila, as he accused the public official and those six expert participants, maintaining that some of the victims changed their minds *due to their ambiguous and therefore negligent public statements*. Then, in 2012, the judge of the Tribunal of L'Aquila, sitting as a sole judge according to the Italian criminal law procedure, also gave the same answer to that question, sentencing the accused – as participants in the supposed negligent communication campaign – to 6 years in prison for manslaughter. In particular, he found the seven defendants guilty of negligence (a) for having assessed the risks related to the seismic activity in an inaccurate, generic and ineffective way, and (b) for having provided the citizenship with imprecise, incomplete and contradictory information, having regard to the nature, the causes, the danger and the future development of the seismic activity (Tribunal of L'Aquila 2012).

In 2014, the Court of Appeal of L'Aquila (court of second instance) absolved the six scientists. The judges stated that according to the most reliable scientific knowledge available in 2009 (and still undisputed) earthquakes are *unpredictable*. Nonetheless, regarding the public official's conduct, the court argued that his declaration (*supra*) was scientifically erroneous and seriously negligent in the light of its possible (foreseeable) effects on the behaviour of the citizens; moreover, it was assumed that his conduct effectively affected (i.e. 'determined') the behaviour of some of the victims. For these reasons, the court confirmed the conviction, reducing the sentence (Court of Appeal of L'Aquila 2014).

Finally, in April 2016, the Italian Supreme Court of Cassation essentially confirmed the sentence of the Court of Appeal (Court of Cassation 2016).

17.4 A Trial Against Science? A Question of Imputation

In the aftermath of the earthquake, the reactions to the beginning of the trial were varied. Initially, many critics (mostly scientists) tried to draw parallels with the Inquisition's trial against Galileo, framing the L'Aquila trial as a 'trial against science' (Hall 2011; Lauta 2014).

Did the judge really convict *Science* as such for asserting that earthquakes are unpredictable? This explication is actually clearly inappropriate (Fornasari and Insolera 2015). The six scientists did not stand accused of ignoring the imminence of an earthquake. The judge himself clearly acknowledged the unpredictable nature of seismic events (Tribunal of L'Aquila 2012: 216). Therefore, no new Galileo trial took place.

So why were the six scientists and the public official put on trial? Perhaps, for what they actually knew (or should have known): that in light of the state-of-the-art scientific knowledge, a *residual risk* of a major earthquake persisted, and this risk was bigger than communicated to the public (Koketsu and Oki 2015). Thus, as everybody already recognised, rather than a trial against science, the first judicial decision convicted the scientists and the public official for their *careless risk communication*.

This interpretation, however, is still problematic. In fact, considering the seriousness of the charge (multiple manslaughters) and according to the generally accepted doctrine in continental criminal law, a 'simple' miscommunication cannot justify *per se* such a conviction (Fornasari and Insolera 2015). Indeed, although risk communication was superficial and ambiguous, this does not automatically mean that it was legally negligent. As we already argued, in order to fairly ascribe harmful events to defendants, each single event (considering all main concurrent etiological factors: victims' decisions to go back home, and the imminent earthquake, in this case) must be considered concretely *ex ante* foreseeable as a result of defendants' conduct (i.e. each given public declaration).

17.5 Science, Policy, and Responsibility

Let us then focus on the two most relevant criticisms concerning the evaluation of the six scientists' role. On the one hand, from the criminal lawyer point of view, the majority of scholars harshly criticised this verdict for failing to carry out the 'but-for' test of causality and for incorrectly deeming defendants to be 'negligent' (e.g. Notaro 2014; Galluccio 2015; Valbonesi 2015). On the other hand, from the wider and interdisciplinary theoretical perspective on the relationships between Law, Science, and Policy, the first judgement, especially, is not convincing due to the misunderstanding of how policy-relevant science should participate in prevention processes and serve society (Amato and Galadini 2013; Simoncini 2014).

Let us start from the second perspective (the wider and interdisciplinary). We shall turn to the first one afterwards (the criminal lawyer point of view).

Scientific knowledge has a relevant, essential normative power. Some authors may refer to the concept of 'regulatory science' (Jasanoff 1995a). We all need science to describe nature, make hypotheses and thereby explain how it might work. This allows us take empirical-based rational decisions. Nonetheless, 'Science cannot tell us how to live. It cannot tell us right and wrong' (Earp 2016: 18). Indeed, although possible interactions among scientific (empirical) and normative (axiological and teleological) dimensions may inform several different models (Tallacchini 2005), there is no doubt that, generally speaking, every political or legal decision-making episode entails carrying out (science-based) *evaluations*. In order to decide how we *ought to behave*, we must possess scientific information but then move from 'mere facts', 'natural laws' and 'scientific hypotheses' (which are frequently controversial and uncertain, after all), to 'models-of-conduct'. Hence, we must turn to values, ethical principles, legal rules, *stare decisis*, political goals, even interests, and any other *normative criteria*.

If this is right, it may explain why the main error made in the earthquake risk's management was the (proven) political decision to use scientists' prestige as an instrument of communication strategy, delegating to the members of the commission the function of reassuring the population. Instead, the necessary interaction between scientists (risk assessors) and policy-makers (risk managers) would have required a clearer definition of roles, tasks, and therefore responsibilities, possibly through the provision of a proper legal framework (Simoncini 2014: 147, 154). Accordingly, scientists should have obtained accurate findings, assessed the risk and presented information about hazards, whereas risk-managers (of the Civil Protection Agency) should have used that information in order to make decisions about public welfare, developing – where possible – feasible responses (Hall 2011, who reports Thomas Jordan's point of view[10]).

Why have scientists been pushed to overstep this limitation? The national agency erroneously evaluated what scientists can and should do in risk prevention processes. Its representatives did not take into account the fundamental distinction drawn between *risk assessment*, on the one side, and *decision-making* (belonging to their jurisdiction), on the other, so they felt free to use scientists' reputations to put on a 'media operation'. Involuntarily accepting this role (instead of just assessing the risk and then reporting its assessment to the Civil Protection Agency), the experts concurred in generating, or reinforcing, an erroneous 'normative expectation', which probably became the 'hidden' motivation of their conviction in the first instance judgement.

This shows that the two criticisms (the two perspectives: the criminal and the wider) are strictly related (Simoncini 2014). As we already argued, and according to the methodological paradigm of 'co-production' (Jasanoff 1995b), the definition of the 'duty of care' – as well as any legal construction (Tallacchini 2004) – depends on how jurists conceive science and therefore turn to scientific knowledge. The judicial construction of such a normative expectation mainly depends on two strictly related factors: (a) the 'image of science' (Villa 1984) that interpreters have in mind, and therefore (b) the role that they assign to science in order to determine normative contents. Indeed, if we conceive science as a tool to solve any challenge, forgetting that decision-making requires turning to certain normative criteria, we will probably tend to delegate to scientists the whole process of risk management and prevention (as it occurred in L'Aquila) and therefore conceive *its failure as due to their fault*.

In our opinion, this also might explain why the tribunal turned to scientific knowledge in such an unusual way. As we already pointed out, both the prosecutor and the first judge declared that they knew that in the light of the state-of-the-art scientific knowledge the seismologists could not foresee the L'Aquila earthquake. So why did they decide to convict them? How could the experts foresee and prevent what happened... *earthquakes being unpredictable*?

We can find several interpretations of the judge's reasoning.

Some scholars, for instance, argue that he retrospectively applied the so-called Precautionary Principle to criminal negligence. We do not entirely agree. We cannot deal here with the issue concerning the role that this principle might play in

[10] Director of the Southern California Earthquake Centre at the University of Southern California in Los Angeles, and chair of the International Commission on Earthquake Forecasting (ICEF).

individual criminal liability.[11] Even so, there is no doubt that the Precautionary Principle (as it is commonly understood, at least) needs a scientific base to be applied – even if this be uncertain, controversial, not yet generally accepted by the scientific community, insufficient or merely hypothetical (Cortina 2004; Romeo Casabona 2004; Tallacchini 2005; O'Mathúna 2011). On the contrary, in the L'Aquila trial no *alternative reliable* scientific hypothesis could allow that those tragic events (the earthquake and each resulting fatal event) were concretely foreseeable (Perin 2014: 1395).

Thus, we may give the following explanation: the overestimation of what scientists can and therefore should do may have induced the judge to assume the criterion according to which 'good' decisions under uncertainty[12] should be taken by imagining, or supposing, the *worst possible scenario*. The Italian Supreme Court of Cassation, after all, already provided a similar solution in an analogous case (a disastrous flood), the so-called 'Sarno case' (Court of Cassation 2010). Perhaps this kind of reasoning represents an adaptation of the 'maximin principle' (Rawls 1971; Ewald 2002), as a criminal scholar suggests (Castronuovo 2012: 32). In any case, it definitely and clearly confirms courts' tendency of judging defendants' conducts in the light of what already happened, *a posteriori*, thereby supporting excessive social *expectations* – no longer reasonable, but omniscient models –, and the resulting desire for compensation and revenge.[13]

By continuing to do so, however, the courts do not suggest 'duties of care' that provide a 'good service to society' (Amato and Galadini 2013: 13). Similar prescriptive content would impose a duty to act in order to prevent not the concretely 'foreseeable' but the 'worst', that is, a possible future horizon that cannot be exactly predetermined.[14] Such a 'normative expectation' can be 'constructed' by a narrative masterpiece, like the one written by Leonardo Sciascia, but not by crimi-

[11] Let us just note the twentieth century normative theory of negligence has been based (by most European Continental criminal law scholars, at least) on a positivist concept of 'scientific law'. This means that only a *definitively confirmed nomological 'base of knowledge'* (i.e. a 'scientific law') would allow consider harmful events *objectively ex ante foreseeable*, and therefore *imputable* (e.g. Forti 1990). Nevertheless, the growing difficulty in distinguishing 'scientific laws' (already confirmed) from 'mere hypotheses' (still under discussion, not yet corroborated) inevitably affects this mainstream normative construction of 'criminal negligence'. Indeed, this 'crisis of the nomological model' (Perin 2014) might also explain the central importance of the 'Precautionary Principle' in the recent academic debate on this field (e.g. Pardy 2002; Romeo Casabona 2004; Roets 2007; Pulitanò 2008; Castronuovo 2012; Perin 2017). *Quid iuris* when the available scientific knowledge is uncertain, i.e. not yet definitely confirmed by the scientific community? Can we establish 'duties to act' – that is, 'duties of care' –, turning to controversial or not yet generally accepted nomological statements?

[12] Not in the sense of 'scientific uncertainty'. Of course, the nature (and the degree) of uncertainty can vary. For instance, there can be uncertainty about the (effectiveness of) measures that could be implemented to reduce, or to prevent certain *possible* (although *not foreseeable*) events (De Jong 2013).

[13] We refer to the so-called *hindsight bias*, which mostly affect medical malpractice verdicts (Peters 2002; Haskel 2007; Oeberst and Goeckenjan 2016).

[14] Indeed, the Tribunal did not clearly determine *how* the defendants ought to have acted.

nal courts. Otherwise, the legal concept of 'negligence' would completely lose the fundamental – although highly discretional and 'pluralistic' (Fletcher 1985; Moran 2003) – reference of the 'reasonable person' normative model, leaving individuals 'entirely alone taking responsibilities for choices that might be tragic' (Castronuovo 2012: 146).[15]

17.6 Why We Blame Experts (Concluding Remarks)

In order to summarise and conclude, we shall briefly suggest three main reasons (not the only reasons, of course) for the recent tendency to blame experts, resulting in an increased need for legal processes.

The first reason is cultural: epistemological, more precisely. We (society) still tend to think that science (empirical, or 'hard', sciences) can enable us to adequately manage any natural or technological risk and successfully respond to any resulting possible crisis. This optimistic, or 'positivist', 'image of science' (Villa 1984) inevitably increases our expectations and desire for revenge (towards scientists as well as experts or professionals such as medical practitioners: think about increasing medical malpractice claims) when something 'bad' or 'wrong' happens, disappointing our hopes.

This old-fashioned and ingenuous portrayal frequently also influences decision-makers (like the Civil Protection Agency) and court judgements, due to the fundamental role that scientific knowledge plays in constructing normative criteria and prescriptive contents. There is a deep 'epistemological relationship' between legal categories (e.g. 'negligence') and scientific knowledge, and this implies that judges, legal scholars and institutions of law enforcement should understand more adequately scientific culture and its language (Pulitanò 2007: 876).

The last reason is the (possible) lack of a proper legal framework. This should distinguish between roles and tasks of scientists (risk assessment) and political bodies (decision-making) in systems or processes of risk prevention (Simoncini 2014). An adequate legal framework should draw a clear accountability regime, limiting the potential responsibility of experts according to the role that science should play in deciding how we *ought to behave*.

Acknowledgments The author would like to thank Ilaria Campagna (PhD, University of Trento) for her insightful and helpful comments. The author also gratefully acknowledges the suggestions of Gabriele Fornasari (Full Professor of Criminal Law, University of Trento) and Dónal O'Mathúna (co-editor).

---

[15] Or, even worse, taking responsibilities for decisions (not) taken by others (again, politicians). We refer to the persistent vulnerability of the buildings destroyed by the earthquake, whose collapse was actually the 'main reason' for the slaughter (Fioritto 2014; Fornasari and Insolera 2015).

References

Amato, Alessandro, and Fabrizio Galadini. 2013. Gli argomenti della scienza nel processo dell'Aquila alla "Commissione Grandi Rischi". *Analysis* 3-4: 2–37.

———. 2015. Media e traduzione delle conoscenze scientifiche prima del terremoto del 2009 (Media translation of scientific knowledge before the 2009 earthquake). In *Terremoti, comunicazione e diritto. Riflessioni sul processo alla "Commissione Grandi Rischi"*, ed. Alessandro Amato, Andrea Cerase, and Fabrizio Galadini, 169–202. Milano: Franco Angeli.

Baron, Marcia. 2012. The standard of the reasonable person in the criminal law. In *The structures of the criminal law*, ed. R.A. Duff, Lindsay Farmer, S.E. Marshall, Massimo Renzo, and Victor Tadros. Oxford: Oxford Scholarship Online. https://doi.org/10.1093/acprof:oso/9780199644315.003.0002.

Bex, Floris, and Douglas Walton. 2012. Burdens and standards of proof for inference to the best explanation: Three case studies. *Law, Probability and Risk* 11: 113–133. https://doi.org/10.1093/lpr/mgs003.

Bicchieri, Cristina. 2014. Norms, conventions, and the power of expectations. In *Philosophy of social science: A new introduction*, ed. Nancy Cartwright and Eleonora Montuschi, 208–229. Oxford: Oxford University Press.

Castronuovo, Donato. 2009. *La colpa penale*. Milano: Giuffrè.

———. 2012. *Principio di precauzione e diritto penale. Paradigmi dell'incertezza nella struttura del reato*. Roma: Aracne.

Chernov, Dmitry, and Didier Sornette. 2016. *Man-made catastrophes and risk information concealment: Case studies of major disasters and human fallibility*. Cham: Springer. https://doi.org/10.1007/978-3-319-24301-6.

Cortina, Adela. 2004. Fundamentos filosóficos del principio de precaución. In *Principio de precaución, Biotecnología y Derecho*, ed. Carlos María Romeo Casabona, 3–16. Bilbao/Granada: Inter-University Chair in Law and the Human Genome – Comares.

Court of Appeal of L'Aquila. 2014. 10.11.2014–6.02.2015, No. 3317. http://www.penalecontemporaneo.it/upload/1423587968CA_LAquila_GrandiRischi.pdf. Accessed 9 Oct 2018.

Court of Cassation, criminal section IV. 2010. 11.3.2010–3.5.2010, No. 16761. http://www.ambientediritto.it/sentenze/2010/Cassazione/Cassazione_2010_n._16761.pdf. Accessed 9 Oct 2018.

Court of Cassation, criminal section IV. 2016. 19.11.2015–25.03.2016, No. 12478. http://www.penalecontemporaneo.it/upload/1459777212Cass_2016_12478_LAquila.pdf. Accessed 9 Oct 2018.

De Jong, Elbert. 2013. Regulating uncertain risks in an innovative society: A liability law perspective. In *Robotik Und Recht I*, ed. Eric Hilgendorf and Jan-Philipp Günther, 163–183. Baden-Baden: Nomos Verlag.

Donini, Massimo. 2006. *Imputazione oggettiva dell'evento. "Nesso di rischio" e responsabilità per fatto proprio*. Torino: Giappichelli.

Douglas, Heather E. 2003. The moral responsibilities of scientists (tensions between autonomy and responsibility). *American Philosophical Quarterly* 40: 59–68.

Douven, Igor. 2017. Abduction. In *The Stanford encyclopedia of philosophy*, ed. Edward N. Zalta, Summer 2017 edition. https://plato.stanford.edu/archives/sum2017/entries/abduction. Accessed 9 Oct 2018.

Earp, Brian D. 2016. Science cannot determine human values. *Think* 15: 17–23. https://doi.org/10.1017/S1477175616000026.

Ewald, François. 2002. The return of Descartes' malicious demon: An outline of a philosophy of precaution. In *Embracing risk: The changing culture of insurance and responsibility*, ed. Tom Baker and Jonathan Simon, 273–301. Chicago: University of Chicago Press.

Fioritto, Alfredo. 2014. Science, scientist and judges: Can judges try science? *European Journal of Risk Regulation* 5: 133–136.

Fletcher, George P. 1971. The theory of criminal negligence: A comparative analysis. *University of Pennsylvania Law Review* 119: 401–438. https://doi.org/10.2307/3311308.

———. 1985. The right and the reasonable. *Harvard Law Review* 98: 949–982. https://doi.org/10.2307/1340881.

Fornasari, Gabriele. 1990. *Il principio di inesigibilità nel diritto penale*. Padova: Cedam.

Fornasari, Gabriele, and Gaetano Insolera. 2015. Scienza, rischi naturali, comunicazione del rischio e responsabilità penale. Il punto di vista del penalista (Science, natural hazards, risk communication and legal accountability: A criminal lawyer point of view). In *Terremoti, comunicazione e diritto. Riflessioni sul processo alla "Commissione Grandi Rischi"*, ed. Alessandro Amato, Andrea Cerase, and Fabrizio Galadini, 7–12. Milano: Franco Angeli.

Forti, Gabrio. 1990. *Colpa ed evento nel diritto penale*. Milano: Giuffrè.

Galimberti, Umberto. 2009. Man in the age of technology. *Journal of Analytical Psychology* 54: 3–17. https://doi.org/10.1111/j.1468-5922.2008.01753.x.

Galluccio, Alessandra. 2015. Comunicazione (scientifica) e responsabilità penale: riflessioni sulla causalità psichica a margine della sentenza "Grandi Rischi" (Scientific communication and criminal liability: Reflections on "psychic" causality on the sidelines of "Major Risks" verdict). In *Terremoti, comunicazione e diritto. Riflessioni sul processo alla "Commissione Grandi Rischi"*, ed. Alessandro Amato, Andrea Cerase, and Fabrizio Galadini, 293–307. Milano: Franco Angeli.

Gimbernat Ordeig, Enrique. 1985-6. ¿Que es la imputación objetiva? *Estudios penales y criminologícos* 10: 167–186.

Hall, Stephen S. 2011. Scientists on trial: At fault? *Nature* 477: 264–269. https://doi.org/10.1038/477264a.

Harris, John. 2014. Science and freedom. *Revista de Derecho y Genoma Humano/Law and the Human Genome Review*, Special Issue: 103–108.

Haskel, Michael A. 2007. A proposal for addressing the effects of hindsight and positive outcome biases in medical malpractice cases. *Tort Trial & Insurance Practice Law Journal* 42 (3): 895–940. http://www.jstor.org/stable/25763860. Accessed 9 Oct 2018.

Holmes, Oliver Wendell, Jr. 1881. *The common law*. Boston: Little, Brown, and Co.

Jasanoff, Sheila. 1995a. Procedural choices in regulatory science. *Technology in Society* 17: 279–293. https://doi.org/10.1016/0160-791X(95)00011-F.

———. 1995b. *Science at the bar: Law, science and technology in America*. Cambridge, MA: Harvard University Press.

Jonas, Hans. 1984. *The imperative of responsibility: In search of ethics for the technological age*. Chicago: University of Chicago Press.

Koketsu, Kazuki, and Satoko Oki. 2015. Il processo dell'Aquila: l'incertezza della scienza dei disastri e le responsabilità degli scienziati (L'Aquila trial: Uncertainty in disaster sciences and scientists' responsibility). In *Terremoti, comunicazione e diritto. Riflessioni sul processo alla "Commissione Grandi Rischi"*, ed. Alessandro Amato, Andrea Cerase, and Fabrizio Galadini, 63–86. Milano: Franco Angeli.

Lauta, Kristian C. 2014. New fault lines? On responsibility and disasters. *European Journal of Risk Regulation* 5: 137–145.

———. 2015. *Disaster law*. Abingdon/New York: Routledge.

Mantegna, Rosario Nunzio. 2006. The tenth article of Ettore Majorana. *Europhysics News* 37: 15–17. https://doi.org/10.1051/epn:2006402.

Marinucci, Giorgio. 1965. *La colpa per inosservanza di leggi*. Milano: Giuffrè.

Melendo Pardos, Mariano. 2002. *El concepto material de culpabilidad y el principio de inexigibilidad. Sobre el nacimiento y evolución de las concepciones normativas*. Granada: Comares.

Moran, Mayo. 2003. *Rethinking the reasonable person: An egalitarian reconstruction of the objective standard*. Oxford: Oxford University Press. https://doi.org/10.1093/acprof:oso/9780199247820.001.0001.

Notaro, Domenico. 2014. Scientists and earthquake risk prediction: "Ordinary" liability in an extraordinary case? *European Journal of Risk Regulation* 5: 159–167.

O'Mathúna, Dónal P. 2011. Taking a precautionary approach to nanotechnology. *Nanotechnology Development* 1: 26–30. https://doi.org/10.4081/nd.2011.e6.
Oeberst, Aileen, and Ingke Goeckenjan. 2016. When being wise after the event results in injustice: Evidence for hindsight bias in judges' negligence assessments. *Psychology, Public Policy, and Law* 22 (3): 271–279. https://doi.org/10.1037/law0000091.
Pardo, Michael S., and Ronald J. Allen. 2008. Juridical proof and the best explanation. *Law and Philosophy* 27: 223–268. https://doi.org/10.1007/s10982-007-9016-4.
Pardy, Bruce. 2002. Applying the precautionary principle to private persons: Should it affect civil and criminal liability? *Les Cahiers de droit* 43: 63–78. https://doi.org/10.7202/043702ar.
Perin, Andrea. 2014. La crisi del "modello nomologico" fra spiegazione e prevedibilità dell'evento nel diritto penale. Note introduttive e questioni preliminari sul fatto tipico colposo (The crisis of the 'nomological model' between explanation and foreseeability in criminal law: Introductory notes and preliminary issues on criminal negligence). *Rivista Italiana di Diritto e Procedura Penale* 57: 1371–1409.
———. 2017. La condotta lesiva colposa. Una prospettiva ricostruttiva (Negligent harmful conduct. A reconstructive perspective). *Política Criminal/Criminal Policy* 12: 207–266. https://doi.org/10.4067/S0718-33992017000100007.
———. 2018. L'imputazione oggettiva dell'evento per omissione impropria. Argomenti a favore della "diminuzione di *chances*". *Archivio Penale* (2 web): 1–47. http://www.archiviopenale.it/limputazione-oggettiva-dellevento-per-omissione-impropria-argomenti-a-favore-della-diminuzione-di-chances/articoli/17467. Accessed 9 Oct 2018.
Peters, Philip G., Jr. 2002. The role of jury in modern malpractice law. *Iowa Law Review* 87: 909–970.
Pulitanò, Domenico. 2007. Il diritto penale fra vincoli di realtà e sapere scientifico. In *Scritti per Federico Stella I*, 851–890. Napoli: Jovene.
———. 2008. Colpa ed evoluzione del sapere scientifico. *Diritto penale e processo* 5: 647–654.
Rawls, John. 1971. *A theory of justice*. Cambridge: Harvard University Press.
Recami, Erasmo. 2006. The scientific work of Ettore Majorana: An introduction. *Electronic Journal of Theoretical Physics* 10: 1–10.
———. 2014. Sciascia e Majorana. Il problema della responsabilità dello scienziato. *Studium* 3: 407–415.
Roets, Damien. 2007. Réflexion sur les possibles implications du principe de précaution en droit pénal de l'imprudence. *Revue de science criminelle et de droit pénal comparé* 2: 251–266.
Romeo Casabona, Carlos M. 2004. Aportaciones del principio de precaución al Derecho Penal. In *Principio de precaución, Biotecnología y Derecho*, ed. Carlos María Romeo Casabona, 385–422. Bilbao/Granada: Inter-University Chair in Law and the Human Genome – Comares.
Roxin, Claus. 1994. *Strafrecht. Allgemeiner Teil, Band I: Grundlagen. Der Aufbau der Verbrechenslehre*. München: Beck.
Scalet, Steven P. 2003. Fitting the people they are meant to serve: Reasonable persons in the American legal system. *Law and Philosophy* 22: 75–110. https://doi.org/10.1023/A:1022540824518.
Sciascia, Leonardo. 1987. *The mystery of Majorana*. New York: Carcanet Press.
Simoncini, Marta. 2014. When science meets responsibility: The major risks commission and the L'Aquila earthquake. *European Journal of Risk Regulation* 5: 146–158.
Tallacchini, Maria Chiara. 2004. La costruzione giuridica dei rischi e la partecipazione del pubblico alle decisioni science-based. In *Scienza e diritto nel prisma del diritto comparato*, ed. Giovanni Comandé and Giulio Ponzanelli, 339–355. Torino: Giappichelli.
———. 2005. Before and beyond the precautionary principle: Epistemology of uncertainty in science and law. *Toxicology and Applied Pharmacology* 207: 645–651. https://doi.org/10.1016/j.taap.2004.12.029.

Tribunal of L'Aquila, criminal section. 2012. 22.10.2012, No. 380. http://www.penalecontemporaneo.it/upload/1358763839SENTENZA%20GRANDI%20RISCHI.pdf. Accessed 9 Oct 2018.

Tuzet, Giovanni. 2006. *La prima inferenza. L'abduzione di C. S. Peirce fra scienza e diritto*. Torino: Giappichelli.

Valbonesi, Cecilia. 2015. Scienza sismica e responsabilità penale. Riflessioni sul rimprovero per colpa a margine del processo dell'Aquila (Seismology and criminal liability: Reflections on the manslaughter allegations on the sidelines of L'Aquila trial). In *Terremoti, comunicazione e diritto. Riflessioni sul processo alla "Commissione Grandi Rischi"*, ed. Alessandro Amato, Andrea Cerase, and Fabrizio Galadini, 265–292. Milano: Franco Angeli.

Villa, Vittorio. 1984. Legal science and hermeneutic point of view. In *Theory of legal science*, ed. Aleksander Peczenik, Lars Lindahl, and Bert Van Roermund, 509–520. Dordrecht: Springer. https://doi.org/10.1007/978-94-009-6481-5_43.

Westen, Peter. 2008. Individualizing the reasonable person in criminal law. *Criminal Law and Philosophy* 2: 137–162. https://doi.org/10.1007/s11572-007-9041-2.

Index

D. P. O'Mathúna, I. de Miguel Beriain (eds.), *Ethics and Law for Chemical, Biological, Radiological, Nuclear & Explosive Crises*, The International Library of Ethics, Law and Technology 20, https://doi.org/10.1007/978-3-030-11977-5

Printed by Printforce, the Netherlands